T0332502

OCEAN RESOURCES

OCEAN RESOURCES

VOLUME I

ASSESSMENT AND UTILISATION

Derived from papers presented at the
First International Ocean Technology Congress
on EEZ Resources: Technology Assessment
held in Honolulu, Hawaii, 22–26 January 1989

edited by

DENNIS A. ARDUS

British Geological Survey, Edinburgh, U.K.

and

MICHAEL A. CHAMP

National Science Foundation, Washington DC, U.S.A.

KLUWER ACADEMIC PUBLISHERS
DORDRECHT / BOSTON / LONDON

Library of Congress Cataloging-in-Publication Data

International Ocean Technology Congress on EEZ Resources: Technology
  Assessment (1st : 1989 : Honolulu, Hawaii)
    Ocean resources / edited by Dennis A. Ardus and Michael A. Champ.
      p.   cm.
    "Derived from papers at the First International Ocean Technology
  Congress on EEZ Resources: Technology Assessment, Honolulu, Hawaii,
  1989."
    Includes index.
    Contents: v. 1. Assessment and utilisation -- v. 2. Subsea work
  systems and technologies.
    ISBN 0-7923-0954-5 (set : alk. paper). -- ISBN 0-7923-0952-9 (v. 1
  : alk. paper). -- ISBN 0-7923-0953-7 (v. 2 : alk. paper)
    1. Marine resources--Congresses. 2. Ocean engineering-
  -Congresses.   I. Ardus, D. A.   II. Champ, Michael A.   III. Title.
  GC1001.I58 1990
  333.91'64--dc20   90-5203

ISBN 0-7923-0952-9(I)
ISBN 0-7923-0953-7(II)
ISBN 0-7923-0954-5(set)

Published by Kluwer Academic Publishers,
P.O. Box 17, 3300 AA Dordrecht, The Netherlands.

Kluwer Academic Publishers incorporates
the publishing programmes of
D. Reidel, Martinus Nijhoff, Dr W. Junk and MTP Press.

Sold and distributed in the U.S.A. and Canada
by Kluwer Academic Publishers,
101 Philip Drive, Norwell, MA 02061, U.S.A.

In all other countries, sold and distributed
by Kluwer Academic Publishers Group,
P.O. Box 322, 3300 AH Dordrecht, The Netherlands.

Printed on acid-free paper

Printed in the Netherlands

DEDICATION

Dr. Kenji Okamura

On behalf of the participants and many friends who attended the first International Ocean Technology Congress (IOTC), we would like to honour posthumously Dr. Kenji Okamura, who was a Special Assistant to the Minister for Science and Technology, a longtime Executive with Mitsubishi Heavy Industries Ltd., and a Founding Director of the Japan Marine Science and Technology Centre.

Dr. Kenji Okamura was internationally known and highly respected for his distinguished career and many contributions to the advancement of ocean science and technology for the development and utilization of the oceans and their resources. Among his many accomplishments was his distinguished service and valuable contributions as participant and advisor to several Marine Technology Panels of the U.S.-Japan Cooperative Program in Natural Resources (UJNR).

Dr. Okamura died on January 15th, 1989, one week before the IOTC in Honolulu, Hawaii. He prepared several papers for this conference which were presented by others and incorporated into the conference record.

In his honour, we would like to dedicate this International Ocean Technology Congress and the resultant conference papers to the memory of Dr. Kenji Okamura.

Dr. Okamura will be remembered for his pursuit of the development of the oceans for the benefit of mankind.

The International Ocean Technology Congress

PREFACE

Today western nations consume annually only a small percentage of their resources from the sea, despite the proclamation of Exclusive Economic Zones (EEZ) by many. In contrast, most Pacific Basin Countries obtain more than a quarter of their annual needs from the ocean. Determination of greater rewards from the development of marine resources is markedly inhibited by the limited technical abilities available to locate and assess them.

Knowledge of Exclusive Economic Zone resources is schematic and generalised, and a detailed understanding of the geology and processes relating to the economic use of the seafloor is both fragmentary and very basic. Technology for mapping the mineral resources of continental shelves and ocean areas, except in active offshore hydrocarbon provinces, has been largely developed in pursuit of scientific objectives and competence to rapidly appraise economic potential is limited. Similarly, the capability to characterise and evaluate the other resources of the seas is rudimentary.

The development of ocean resources will become increasingly urgent as the growth of the world population and the depletion of land reserves combine to enhance demand. Also, increasing environmental constraints will limit the availability of traditional land-based resources; nevertheless, new offshore development must proceed in a manner whereby the marine environment is not plundered but protected and conserved. The challenge to develop ocean resources with responsible environmental stewardship will require greater leadership than the development of the technologies of exploitation.

The papers in this volume have been derived from those presented at the International Ocean Technology Congress on EEZ resources - Technology Assessment held in Hawaii in 1989. They demonstrate a breadth of developing awareness and capabilities, concepts and constraints. These suggest that exploitation of the many and varied opportunities will make the oceans a major activity area given the necessary statutory and regulatory framework needed for orderly, responsible and economic development. However, a database of information is necessary, not only to provide awareness of resources, but as a baseline to allow a balanced view of environmental factors and economic benefits to be considered when licencing and management strategies are developed by nations.

Extensive mining of marine hard mineral resources is presently restricted to aggregates and placer minerals; while recovery of phosphates, polymetallic sulphides and cobalt, manganese and nickel enriched crusts and nodules remains academic in economic terms. Alongside exploration and evaluation there is a parallel requirement for geotechnical data and awareness of potential geohazards in order that investigations, extraction procedures or the installation of structures can proceed with the minimum of risk.

In the context of world demographics and rising expectations, energy demand is anticipated to grow by a factor of ten over the next century even assuming extensive conservation and limited population growth. These constraints are not evident at present and renewable energy presents diverse opportunities dependant on geographic location,

geological setting and oceanographic regime. Considerable promise is
given by wave power, tidal power and by Ocean Thermal Energy Conversion
(OTEC).

The benefits of employing deep ocean water as a nutrient medium is
seen to complement OTEC systems. Improved efficiencies are enhanced by
the integration of energy production with various mariculture and
agriculture operations. Deep ocean water stimulates high productivity,
as in areas of natural upwelling, with its biological purity, freedom
from pathogens and competitor organisms, and richness in nutrients
which allow production of marine protein in a manner more reminiscent
of a controlled manufacturing process than farming.

The utilisation of ocean space also provides a resource simply by
direct land extension through reclamation or by the establishment of
sea-borne communities nearshore to alleviate the pressures evident in
many coastal metropoli. Alternatively, and more futuristically,
theoretically self-sufficient communities, offshore airports and
recreational facilities (hotels, theme parks, etc.) could be
established in oceanic waters.

Additionally, the oceans role as an assimilator of selected waste
materials, in the face of increasing environmental difficulties
onshore, needs to be appraised with a full awareness of the physical,
chemical and biological processes in the marine environment, with their
temporal and spatial scales, and subsequent cause and effect
relationships in marine pollution. Coastal waters in many areas are
showing acute sensitivity to the level of contaminents and integrated
monitoring of onshore drainage basins, the adjacent coastal zones and
the seas beyond is required. It is possible that the discharge of
these wastes into more open, deeper oceanic waters would provide
greater dilution and subsequently render contaminants discharged to
background levels without deleterious effects.

When we consider space, within our solar system only the earth
has an atmosphere and ocean to support life. Therefore, the
exploration, protection and development of ocean space and it´s
resources must be given a major priority by mankind.

                                        Dennis A. Ardus and Michael A. Champ

FOREWORD

Five billion men, women and children live on planet Earth and the number is increasing rapidly as we strive, with varying degrees of success, to conquer starvation, disease and war; before the middle of the next century the number could be ten billion. Indeed, most of the people who have lived since the beginning of civilisation are alive today. Some of us live very well but most are poor and many are in poverty while some are starving. We all strive for better standards.

To support our lifestyle we need food, minerals and energy nearly all of which we get from a small proportion of the land mass much of which is overworked, overexploited, overfarmed and overpopulated. Mineral recovery despoils the landscape; energy from fossil fuel despoils the atmosphere and threatens global climate, while nuclear power at least for the present has become suspect; food production by intensive arable and livestock farming is beginning to reveal its downside with mounting problems ahead. We release wastes into the atmosphere and the sea and this has to stop. We must take greater care in mining, be less profligate in the burning of fossil fuels which is disturbing the equilibrium of the atmosphere which, in turn, protects us from extremes of heat and cold, and we must heed the warnings of too intensively farming the land, but that is not all. We need to find other sources of these essential resources; and we need more space.

Increasingly we are becoming aware of the scale and importance to us of the oceans, how little we know about them, how they circulate and affect our climate, what they contain in the water column and what they cover on and below the sea bed, and how they can contribute to our insatiable drive to improve our living standards while the world's population grows.

The papers in this volume are concerned with these resources in the Exclusive Economic Zone, that part of the ocean lying within 200 miles of land. This covers about a third of the surface of the Earth and has an area as large as the total land mass itself including the arid deserts and the frozen wastes; yet half of it is over a mile deep and out of our reach today for all but limited exploration. Who knows what knowledge and riches this exploration will reveal.

These papers review first the legal regime, of vital importance for more than territorial reasons since all nations share a common responsibility to protect the oceans from indiscriminate dumping and overexploitation. The old habits of the developed nations will have to change and it will not be easy; but change them we must if we are to convince those less fortunate than we are that these practices must be controlled. Successive chapters assess our geological knowledge and the gaps to be filled to give us a better understanding of the geological processes; the minerals available for extraction; the ocean renewable energy; the living resources; and the utilisation of the ocean space and the opportunities this will present. The final chapters review the environmental assessment which links back to the legal regime.

Together they represent the state of present knowledge and project the opportunities ahead for us. There is much to be learned and even more to be done. The subject is as vast as the oceans themselves,

involving virtually all the sciences and most of the technologies. The
outcome will have a profound effect on living standards in the future
and will ultimately determine just how many people this planet of ours
can support. It affects, and therefore concerns, everyone. It
promises also to be an exciting venture of exploration, discovery, and
implementation of the knowledge acquired to the benefit of all of us.
We can spoil our ocean regime if we disregard the consequences of our
actions. We can turn it into a sterile, poisoned lake unable to
support marine life and harmful to all who come into close contact with
it. By misuse we can modify its interaction with the air to control
our climate with grave consequences.

While the papers in this volume may be dominated by authors from
the USA and UK, they are a truly international set and this is
important because these issues cannot be faced in isolation.

The Congress at which they were presented had three main purposes.
First, to provide an international forum for countries to inform each
other of planned ocean activities and policies; second, to create a
supportive environment in which ocean projects requiring multi-nation
co-operation and/or funding can be proposed; and lastly, to advocate
the development of ocean resources in an environmentally-acceptable
manner. These papers show that these objectives have been partly met
and a step forward in understanding has been made. But a long road
lies ahead and we must rise to the challenge and keep moving forward.
To do this we must keep our imaginations intact and think the
unthinkable for the future will be different in many ways from that
which we can possibly predict today.

A. Gordon Senior
Gordon Senior Associates
Normandy
Surrey
U.K.

ACKNOWLEDGEMENTS

The need for the International Ocean Technology Congress (IOTC) was recognised at a National Science Foundation and University of Hawaii sponsored conference in 1986 concerned with 'Engineering solutions for the Utilization of the Exclusive Economic Zone (EEZ) Resources'. This resulted in the establishment of a small international group of scientists and engineers with a common interest in the development and conservation of ocean space and resources which met at Heriot-Watt University, Edinburgh in 1987. Subsequently, the group has held planning meetings at the Energy and Mineral Research Organization in Taiwan and at the University of Hawaii in 1988.

The Congress, from which these papers were derived, was held in January 1989 in Honolulu, Hawaii. It was sponsored by:

National Science Foundation
Commission of European Communities
Institut Francais de Recherche pour l'Exploitation de la Mer
Industrial Technology Research Institute, Taiwan
University of Hawaii
Heriot-Watt University
Society for Underwater Technology
Marine Technology Society

Members of IOTC who have served as members of the IOTC editorial board for this volume include:

Dennis A. Ardus	British Geological Survey, U.K.
Norman Caplan	National Science Foundation, U.S.A.
Michael A. Champ	Environmental Systems Development Inc., U.S.A.
Chen-Tung A. Chen	National Sun Yat-Sen University, Taiwan
John P. Craven	Law of the Sea Institute, University of Hawaii, U.S.A.
Robin M. Dunbar	Heriot-Watt University, U.K.
Michel Gauthier	Institut Francais de Recherche pour l'Exploitation de la Mer, France
Jorgen Lexander	Swedish Defence Research Establishment, Sweden
C.Y. Li	Advisor on Science & Technology, The Executive Yuan, Taiwan
Kenji Okamura	Ministry of State, Science & Technology Agency, Japan
Boris Winterhalter	Geological Survey of Finland
Paul C. Yuen	University of Hawaii, U.S.A.

The considerable contribution of Fay Horie and Carrie Matsuzaki of the University of Hawaii in the organization of the first IOTC and in the preparation of this volume is gratefully acknowledged.
Pamela Pendreigh and Fiona Samson of Heriot-Watt University, Edinburgh are thanked for their preparation of the text for this Volume.

TABLE OF CONTENTS

PART I

LEGAL REGIME

THE IMPACT OF THE LAW OF THE SEA ON OCEAN RESOURCE DEVELOPMENT AND
OCEAN RESOURCE TECHNOLOGY

JOHN PINNA CRAVEN
University of Hawaii
Honolulu,
Hawaii,
U.S.A.

ABSTRACT. Since 1958 there has been a dramatic and continuous change
in the International Law of the Sea and in particular the Law of the
Sea as it relates to Ocean Resources and Ocean Technology. These
changes in the Law of the Sea were based on predictions and
mispredictions by technologists as to the future of resource related
technology. The author has been involved in this prediction process in
public fora since 1958 and in print since 1966. The time scales of
these predictions were made in terms of decades and as a result these
predictions and those of his colleagues have already come home to
roost. Rationalisation now suggests that the technologists of that
decade were not altogether wrong even though the legal community seized
upon technical mispredictions and misconceptions as the basis for
negotiation of substantial elements of the Law of the Sea. At the
first annual meeting of the Law of the Sea Institute oceanic
technologist Willard Bascomb stated, "I should perhaps note that in our
search for underseas minerals we began at the top of the scale with
diamonds and then we descended through platinum and gold, tungsten, tin
and we are getting down to the lower levels now. I think we may never
get down to manganese nodules". At the second annual conference of the
Institute John Mero (he of deep seabed minerals fame) gave equal
prominence to ocean thermal energy (and it´s associated artificial
upwelling) and the manganese nodules. He predicted economic viability
for one or both in from "ten to twenty years" (i.e. 1976-1986).
Students of fact and logic will agree that it is correct to say that
both Bascomb and Mero were right.
     The community prediction however focussed on manganese nodules and
disregarded Deep Ocean Water as the major resource of the open ocean.
The legal effect of this misprediction is thoroughly discussed in a
previous paper and is not relevant here except as it sets the stage
from which we can courageously and rashly make another attempt at
predicting the future interaction of the technology of the sea with the
Law of the Sea.
     Our point of departure, the current status of the law is as clear
and well defined as International Law has ever been. For one reason or
another nearly every nation of the world including the United States
acknowledges that with the exception of Part XI, the words of the

3

D. A. Ardus and M. A. Champ (eds.), Ocean Resources, Vol. I, 3–10.
© 1990 Kluwer Academic Publishers. Printed in the Netherlands.

UNCLOS III text are a definitive statement of the international law of
the sea. If it is true, as this paper hypothesizes, that much of this
nearly universal consensus code is incompatible with the technology of
today and of the near future, then it would not be rash to suggest that
this universal text will require revision. If so it is most
appropriate then to examine the current state of technology and the
developments in technology which now appear to hold the most promise
for humankind.

1.   THE EXCLUSIVE ECONOMIC ZONE

Although the Treaty redefined the regime of the Territorial Sea, the
Contiguous Zone, the Continental Shelf, the Seabed beyond the area of
national jurisdiction and the High Seas, it is in the Exclusive
Economic Zone that technology will play the most dramatic role in the
modification of the nature of the envisioned regime. Drafters of the
convention believed, with good reason, that the primary resource of the
EEZ was fish and that the regulation of conventional fishing,
conservation and allocation of resources was the primary substance of
the regime. As against this prediction it has been the perception of
Dr. Alistair Johnson of the Marconi Corporation that the primary
resource of the ocean is the deep ocean water itself and it has been
the perception of the Japanese as expressed by the late Dr. Kenji
Okamura that an equally important resource is ocean space and ocean
space utilization.
     We should then examine these new predictions in light of the
physics of the ocean and the needs of society. We now see that the
primary resource characteristic of deep ocean water is that it is cold.
Of nearly equal importance is that it is rich in nutrients and that it
is biologically pure (only a few diatoms). To understand the
fundamental significance of the value of cold we should first note that
the earth's weather and climate is the product of a giant natural ocean
thermal energy plant. The solar collector is the tropical ocean in
which a belt of warm surface waters girdles the earth. The cold water
of this natural OTEC machine is the arctic and antarctic surface
waters. The rotation of the earth and the annual variation of the
solar track produce the oceanic circulations which carry the warm water
away from the equator and the cold water away from the arctic and
antarctic zones. The mechanical energy which is extracted from this
thermal process appears in the winds and subsequently the waves and in
the elevation and evaporation of the surface water in the production of
rain and in the production of ocean currents which themselves are
transformed into different kinds of mechanical, chemical and heat
energy. The advantage of the man made version of this natural heat
engine is that the cold water is separated from the warm by a few
thousand feet versus a few thousand miles. As a first result it should
be possible to create microclimate simulations of the earth. We know
this to be the case for the micro-climates associated with the areas of
natural upwelling.
     In utilizing this feature of the deep ocean cold water we should
understand the fundamental nature of natural productivity in any given

micro climate. Each sub region of the earth operates as a heat engine
with transports of heat coming from the solar irradiation, convection
through the transports of fluids such as the gases of the atmosphere
and of the water in the form of atmospheric moisture, river flow,
oceanic transport. At any moment in time, however, the efficiency of
this heat engine is given by the Carnot efficiency $T_1 - T_2/T_1$ where T_1
is the temperature of the warm fluid that participates the thermal
process and T_2 is the temperature of the cold fluid. In desert regions
where the air and ground are at the same temperature T_1-T_2 is very
small and the region is unproductive. In temperate zones in the spring
time when the warm spring sunshine interacts with the cold runoff from
the mountain snows T_1-T_2 is large and as a result there is evaporation,
condensation and rain. The lay public then attributes the remainder of
the growth process to photosynthesis when in point of fact the
thermodynamic processes are equally important. Without an efficient
thermodynamic machine it is not possible to efficiently construct the
high energy molecules which constitute the sugars, starches and oils of
any plant.
     In the low islands of the Pacific subject to trade winds the
temperature does not decrease with altitude until an altitude of about
10,000 feet is reached. As a consequence condensation of moisture does
not occur until the top of the resulting inversion layer is reached
thereby preventing the formation of rain. The isothermal character of
the land mass and the adjacent fluids (land and water) is such that
there is little natural productivity. On the other hand when there are
high islands and winds such as trade winds the moist atmosphere is
lifted into colder regions and "orographic" rain is produced. The
production of rain is a mechanical process in which the heat energy of
the environment is converted to mechanical energy in order to raise the
water from the ocean to the top of the mountain. Once again T_1-T_2 is
large as a result of the vertical transport of the moist atmosphere.
     If now we produce artificial upwelling from the deep ocean to the
surface we dramatically change the temperature differences available to
the natural heat engine. In regions where the water has upwelled the
natural productive efficiency of a low island is increased by factors
of four or more, on the high islands the leeward coasts are similarly
benefitted and in the open sea where there is no natural upwelling an
environmental energy potential now exists in what is now a vast
tropical oceanic desert. The energy required to bring the water to the
surface is, of course minimal, requiring only the energy associated
with the density difference resulting from the difference in
temperature and salinity between the deep ocean and the surface of the
sea. (Indeed the primary energy cost is in the positioning of the deep
water in header tanks and reservoirs from which it may flow by gravity
to the various facilities).
     This realization of the fundamental change in productivity is the
cumulative understanding of a wide variety of deep ocean water
developments that are taking place at the Natural Energy Laboratory in
Ke-ahole point Hawaii. We now understand, that with intelligent
intervention it is possible to achieve the natural benefits of eternal
spring in the tropical oceanic regions. This means the year round
production of spring crops such as strawberries, lettuce, asparagus,

alpine ornamental flowers etc., the year round production at maximum
growth rates of seaweeds such as nori and ogo, of shell fish such as
opihi, oysters, lobster, shrimp, the year round production of kelp of
abalone, of trout, salmon, sea urchins, the year round high volume
production of sophisticated algae such as spirulina, dananiella,
icosopentane, the generation of closed cycle electrical energy without
risk of biofouling and with the use of low cost aluminium heat
exchangers, the flash evaporation of surface water and condensation
with the use of deep ocean water to produce fresh water as a by product
of the open cycle process, the low cost, non-heat producing air
conditioning of buildings, the elimination of chill water generators
and cooling devices in industrial production. Many of these processes
can employ deep ocean water which has already been used or which is yet
to be used.

This technology is being followed closely in Japan and in Europe
and in Britain and Canada. We can confidently predict that various
forms of energy, aquaculture plants will be available for the tropical
islands in the near future. But the major significance of these
developments is the hastening of the transition from fishing to
aquafarming and ranching. This transition is occurring most rapidly in
Japan where aquaculture now provides about twenty five percent of the
marine protein. The construction of artificial reefs is continuing at
a high rate and projects for fertilizing these reefs with deep ocean
water are on the drawing boards. The cage culture of salmon in the
Norwegian fjords is already a major element in the worlds supply of
salmon. The ability to produce marine protein continuously throughout
the year under controlled conditions equivalent to those for the
production of chickens and beef at a cost which is comparable to the
cost of animal protein of comparable quality will displace the
competitive fished product which is the victim of seasonal and annual
variations, of the uncertainty of the hunt, of illegal competitors, of
the cost of regulation, of the economics of quotas and all of the
international encumbrances which make fishing such an interesting and
non-profitable operation.

Of greatest significance to the law of the sea is the elimination
of the need to patrol large areas of the EEZ for the regulation of the
industry. The only fish which has not yet been demonstrated as
amenable to aquaculture is tuna. The American love affair with the
tuna fish sandwich is not likely to be assuaged by a different species
no matter how carefully marinated in mayonnaise, but the first marine
biologist who solves the problem of tuna spawn in captivity will be on
his or her way to fame as the Marine McDonald. It is equally true that
the successful deep ocean water mariculture operations will take place
in the EEZ. The ranching of bottom fish (or slow growing animals such
as lobster) will be most economically conducted in a natural
environment enhanced by artificial reefs and habitats. Enforcement in
these local areas will not be dissimilar to enforcement on the land
ranch.

The pattern for the systematic use of deep ocean water as a
resource is now clear. We can envision a hotel complex in a previously
dry and sterile low island environment. Deep ocean water is brought to
the surface at a temperature of about 6°C. The water is employed in a

closed cycle thermal energy plant for the production of energy adequate
for the hotel complex and the community and industry associated
therewith. The waste cold water from this process now at a temperature
of about 9°C. (Additional cold water may have to be employed to lower
the waste water to this temperature). The water is then employed in
heat exchangers for the air conditioning of the hotel complex, for
chilling fresh water, for industrial cooling and chilling, and for
changing the ground temperature of farm land for the cultivation of
spring crops. The energy saved per unit volume of cold water in
cooling will be an order of magnitude greater than the energy which has
been created. The waste water from these processes will still be pure
and rich in nutrients and can be employed for the high quality marine
products which will characterize the hotel cuisine.

Some decades in the future it will be recognised, as Avery of the
Johns Hopkins Applied Physics Laboratory has already recognised, that
floating OTEC plants can produce ammonia or methanol at costs
competitive with oil. This OTEC process will produce little if any CO_2
in the manufacture of methanol. The greenhouse proof advantages of
these fuel sources will add a considerable bonus to the economic
advantages.

In both short and long term developments, proximity to Islands for
logistic support, for processing and for transport will be an economic
necessity as will the need for an Exclusive Economic Zone for the
management of these ocean complexes. It is in the regime of islands
that the Law of the Sea failed to anticipate the rapid pace of
technology. In the 1958 Convention on the Continental Shelf the width
of the shelf was defined by that famous phrase "to a depth of 200
meters or to a depth which admits of practicable exploitation". The
ink was hardly dry before technology made the depth of practicable
exploitation the bottom of the sea. In a similar manner a soon to be
equally famous phrase of Article 121 states that "Rocks which cannot
sustain human habitation or economic life of their own shall have no
exclusive economic zone or continental shelf". Just as the Continental
shelf phrase was suggestive of a depth limit of 200 meters associated
with the then existing technology of oil drilling and the limit of
economically significant demersal fisheries so the "economic life"
provision suggests a minimum size of rock associated with the technical
feasibility of extracting economic income from a remote location in the
middle of the sea. Lawyers, no doubt have had extensive experience
since the middle of the 19th century with the legal status of "guano
islands". Here the low cost per unit cube of this valuable, for its
time, high tech commodity of modern agriculture suggests a rock like
Nauru which has an area of eight square miles. Nature was not so
instructive to the lawyers as to produce a diamond generating volcanic
cone which produces Hope diamonds on the top of a ten square foot
pinnacle, but technology will be. The most recent development in this
regard is the island of Okino Torishima. This island, which has been
described as a "king size bed" rock sitting on a 2 kilometer by five
kilometer reef is located in a portion of the South Pacific not yet
covered by an EEZ. The Japanese government has sent an expedition of
planners to this island to determine the manner in which it could be
developed for economic purposes. The expressed intent is to include

this island when it declares its 200 mile economic zone. A primary
element of the analysis is the generation of an activity to be
conducted on the rock which will qualify as "capable of sustaining
economic life of its own". For the past few months the rock has been
the locus for a remotely operated weather station. Other possibilities
are as a transfer and repeater station for fibre optic oceanic cables,
as a communications, command and control centre for operations in the
EEZ. The development of a fishery associated with an artificial reef
together with an ocean thermal energy station are other possibilities
under investigation. In any event the Japanese intend to make a
considerable capital investment and to act in a manner which they deem
consistent with the treaty for the establishment of an exclusive
economic zone.

The motivation for this forced development of an accident of
nature is another technological anachronism of the treaty para 8
Article 60. "Artificial islands installations and structure do not
possess the status of islands. They have no territorial sea of their
own and their presence does not affect the delineation of the
territorial sea, the exclusive economic zone or the continental shelf".

The technology for the creation of artificial islands which are
geophysically indistinguishable from natural islands is now well
developed. A major portion of the Netherlands consists of lands
reclaimed from the sea including artificial islands in the North Sea.
The stabilization of the Bangladesh delta will require similar
developments in the process of which artificial and natural islands
will be created by man and by nature as a result of the works of man.
These islands will, for environmental, social, cultural and economic
reasons desire to utilize the ocean space surrounding them at distances
in excess of those required as a safety zone. Although the current
scale of technology will rarely suggest that this zone of ocean space
utilization will be as large as 200 miles, there are many current
examples (the siting of ocean thermal energy facilities, scientific
stations like project DUMAND, artificial reefs for fish habitats,
underwater parks etc.) in which the distances from the port island are
of the order of ten to fifty miles. But ocean space utilization does
not require the building of an artificial island as its central urban
city. The study of stable ocean platforms for urban systems is now
more than fifty years old (Armstrong Seadrome 1932, Mulberry Harbours
1942, Triton City 1965, Aquapolis 1974, Ocean Information City 1985).
The initial realization is that of Australia's Barrier Reef Hotel.
This is a major hotel complex which are based on floating barges above
the barrier reef some 30 kilometres beyond the territorial sea in
Australia's exclusive economic zone. Located in a reef lagoon the
natural protection against sea states which is thus afforded permits
the use of barge construction for the facilities. These include
recreational tennis courts mounted on a separate barge facility and a
flotilla of recreational craft to carry visitors to the coral reef
sites of the Great Barrier Reef Park. Although the concept of a
resident population has not yet been established for this community,
legislation which provides for municipal governance has already been
enacted. The first embryonic elements of other recreational complexes
are now appearing in the form of an underwater hotel in the Carribean

and in the proliferation of tourist submersibles in the Carribean and in Hawaii. Appropriate sites for these submersibles are currently within the territorial sea, but many underwater sites of interest are located well outside of this zone.

This concept of ocean space utilization is being most actively pursued in Japan under both private and governmental auspices. Competition exists between the "Artificial Islands" community and the "Floating Platform" community. The latter is evolving from the stable ship concept. The SWATH (Small Waterplane Twin Hull) or SSP (Semi-Submerged Platform) ship has been conceptually available since the mid fifties but for many years was realized only in the form of the Congressionally unauthorized Kaimalino which achieved legal status as an experimental mobile platform and not as a ship of the United States Navy. Its offspring have been slow in gestation due to cultural lag on the part of rugged seagoing oceanographers and naval mariners who relish the challenge of man against the sea.

Nonetheless the Japanese have successfully developed and commercialized a 400 passenger ferry, an oceanographic ship, an ocean engineering development and submersible support ship and an executive yacht. The chase boat for the America's cup trials was a San Diego SWATH (originally the Soave Lineo). At long last the United States Navy and the Coast Guard have SWATH ships under construction and a sophisticated tourist SWATH ship for Hawaii is presently under construction. This steady unspectacular but continuous progress in the development of stable transportation and stable platforms for ocean space utilization virtually assures the existence of a number of floating communities with significant resident populations in the next two decades. For purposes of taxation, regulation of commerce and trade, participation in the political process of the flag state, suppression of illegal activities, public health and safety, the full emoluments of governance must exist for these communities on the platforms and in the area of ocean space which is to be utilized. The generation of oasis of municipal sovereignty, which include areas of ocean space in the Exclusive Economic Zone, is a societal necessity of these technological advances. Concomitantly there will be a retreat from the perceived necessity to control the full area of the Exclusive Economic Zone and large areas of these legally defined areas of the ocean will return to the status or continue in the status of "res nullius de facto".

2.   CONCLUSION

The actual exploitation of the new regime of the EEZ may come faster than any at this Congress can now anticipate. If it does, the primary exploitation will not be in the form of the minerals of the seabed and the exploitation of deep ocean and Arctic oil. It will be in the form of floating complexes - hotels, theme parks, trade and manufacturing zones, artificial islands, floating and quasi-floating communities. Soon these communities will be self sufficient with deep ocean water as the resource for energy production or energy conservation in all its forms, as the resource for fresh water and as the resource for marine

and vegetable protein of every variety and description. Skeptics
should recall the pace of development of the enclosed shopping mall -
from the first daring innovation in Minneapolis some thirty five years
ago to some 30,000 today. There is already a floating hotel complex in
the Australian EEZ.

PART II

GEOLOGICAL ASSESSMENT

THE US EEZ PROGRAM: INFORMATION AND TECHNOLOGY NEEDS

MILLINGTON LOCKWOOD AND GARY W. HILL
USGS-NOAA Joint Office for Mapping & Research in the EEZ
US Geological Survey
National Center - MS 915
Reston, Virginia 22092, USA

ABSTRACT. In 1983 the US Geological Survey (USGS) and the National Oceanic & Atmospheric Administration (NOAA) initiated a program to characterize the seafloor of the US Exclusive Economic Zone. In the first 5 years of the program over 1,500,000 square nautical miles of the EEZ have been mapped with the GLORIA sidescan sonar system to identify major seafloor features. Results of these surveys are available as atlases and in digital form for advanced image processing.

In addition to the GLORIA surveys, selected areas have been surveyed with the high resolution swath technology available aboard NOAA's research ships. The results are used to produce high-quality bathymetric maps. Since 1984, nearly 60,000 square nautical miles of seafloor have been mapped. The results of these surveys show features and characteristics of the seafloor previously unknown.

This information is pertinent and timely relative to current national policy discussions concerning national/economic security, multiple use decisions, negotiations of international boundaries, ocean waste disposal, marine seafloor research and a range of environmental issues.

1. INTRODUCTION

In 1983, the President proclaimed the establishment of an Exclusive Economic Zone (EEZ), extending 200 nautical miles seaward of the United States coastline. This proclamation extended the Nation's sovereign rights for the purposes of exploring, exploiting, conserving, and managing natural resources in the coastal ocean. This area, approximately 3.4 million square nautical miles is equivalent to 1.3 times the Nation's total land area. Information is needed to determine the resources within it. The Department of the Interior (DOI) and the Department of Commerce (DOC) have determined that the determination of the characteristics and resources of the EEZ is a national priority.

Seafloor bathymetric and imagery maps and data bases with the accuracy and quality being produced by surveys of the EEZ are essential for planning and carrying out resource exploration, exploitation and management activities. Existing maps do not provide adequate feature

D. A. Ardus and M. A. Champ (eds.), Ocean Resources, Vol. I, 13–20.

definition to meet academic, industrial, and government user needs.
The maps are essential to any future commercial development and are
needed by the Nation's scientific community to understand the processes
that form continental margins and the mineral deposits in and on the
seafloor. Advances in computer technology, the development of
multibeam sonar systems, and improved methods of obtaining accurate
marine positions make it possible to respond to these needs.

   In 1984, NOAA and the USGS signed a Memorandum of Understanding to
map the EEZ. Priority areas were identified on the basis of scientific
interest and resource potential - hard mineral as well as oil and gas.
In December 1987, DOI and DOC signed a charter to coordinate mapping
and research in the EEZ. The charter formally establishes a joint
office of the two agencies to coordinate EEZ mapping and research
activities. Additionally, the charter stipulates that coordination
with other federal, state, private and academic organizations having an
interest in EEZ mapping and research is necessary.

   The Joint Office for Mapping and Research in the EEZ (JOMAR) will
provide leadership for the design, implementation and coordination of a
national program and investigation of the non-living resources of the
EEZ seafloor. It will also ensure participation by all interested
groups in the formulation of goals, objectives and priorities for a
national EEZ mapping and research program. Figure 1 shows a schematic
representation of the JOMAR. One of the main functions of the office
will be to develop a 10-year plan for mapping and research in the EEZ
based upon an assessment of "user needs and priorities" accompanied by
an assessment of available technology. This paper outlines the
approach to be taken in order to assure that the program carries out
this mission.

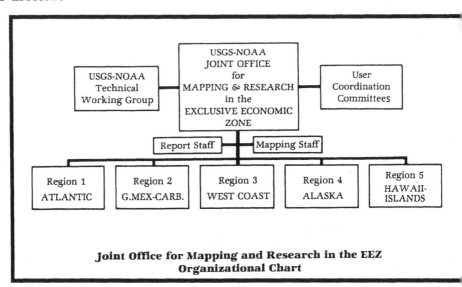

Figure 1. Joint office for mapping and research in the EEZ -
Organizational Chart.

2.   NEEDS OF USERS AND PRODUCTS

Several attempts have been made to assess the data and information requirements of users of the ocean floor. The National Academy of Sciences' Marine Board, the Congressional Office of Technology Assessment and the National Advisory Committee on Oceans & Atmosphere have all conducted evaluations of users of the seafloor. USGS and NOAA have had 3 National Symposia for the purposes of determining applications and uses of seafloor information.

The most recent EEZ Symposium held in November 1987 resulted in approximately 160 recommendations for seafloor information, surveys and maps. These have been evaluated and categorized as follows:

Mapping and Research

1.   Reconnaissance scale data sets and maps are an essential element of ocean exploration.

2.   An understanding of the seafloor in three dimensions using a "corridor" approach should be initiated.

3.   Coastal areas, including state waters, must be part of a national research and mapping strategy.

4.   Selected target or ground truth surveys should be conducted as a means of verifying existing seafloor imagery data.

Technology Development and Implementation

1.   A standard suite of geophysical sensors should be on all EEZ survey ships.

2.   An intermediate scale reconnaissance survey technique should be developed in order to efficiently survey shallow water, continental shelf regions of the EEZ.

3.   Development should be undertaken of an efficient hard substrate drill suitable for use in water depths as great as 4,000 meters.

4.   The Global Positioning System will ensure that positional accuracy is adequate for registration of all data sets.

3.   USERS OF SEAFLOOR MAPS

An indication of the interest in seafloor information is shown by the number of requests for maps and related products from USGS and NOAA distribution centers. For example, in 1985 approximately 250,000 maps showing seafloor features and other marine geological or geophysical information were sold through NOAA and USGS. These include NOAA/USGS "Topo-Bathy" maps as well as conventional bathymetric maps. More than 2,000 requests for bathymetric information were received and over 700 digital bathymetric data sets or graphical products portraying bathymetry were distributed by NOAA's National Geophysical Data Center (NGDC) between 1984-86. Maps which show outer continental shelf lease blocks are routinely produced in cooperation with the Department of the Interior's Minerals Management Service to assist in planning for oil and gas development and the assessment of geological hazards. These maps are an important element of marine environmental impact studies and are used extensively by the offshore marine geophysical exploration industry.

Ocean engineering firms routinely use bathymetric maps for planning, siting of offshore facilities (e.g. offshore structures, energy facilities, terminals), pipeline and telecommunication cable routing, and to support submersible and deep tow surveys. For living resource applications, commercial and recreational fishermen rely on these maps to locate fish habitat areas and delineate seafloor topography for deep draggers and others concerned with bottom fishing and trawling. When enhanced by the addition of bottom sediment information and other environmental data, seafloor maps serve as a valuable tool for marine assessments. Table 1 summarizes the most common uses for seafloor maps.

TABLE 1. Seafloor map uses

- Mineral exploration and development - oil and gas, hard minerals
- Deployment of research instrumentation on or near the seafloor
- Submersible operations
- Fishing operations using deep trawl or bottom fishing gear
- Subsea pipeline or cable routing
- Geological hazard assessments
- Marine geophysical research
- Surveys of ocean waste disposal sites and other environmental impact applications
- Fish habitat research investigations

---

4.   MAPPING TECHNOLOGY DEVELOPMENTS

In the 1950's and 1960's, the technology for seafloor mapping was upgraded through improvements in navigational positioning, i.e. LORAN A and C, Transit satellite navigation, performance and capabilities of survey ships, and enhancements in sounding systems due to advances in electronics and computer processing.

These improvements produced more accurate soundings. However, to provide detailed coverage of a particular area, it was necessary to reduce the distance between survey lines so that all major features were detected. Generally, once an unusual feature was detected or anticipated, the ship ran crossing or perpendicular survey lines to further develop the topography. The development of the sidescan sonar and multibeam technology was a direct response to these needs and a more efficient means to define topography.

5.   TECHNOLOGIES FOR EXPLORATION OF THE SEAFLOOR

A wide range of technological capabilities for exploring the seabed are
presently available and in use. These range from reconnaissance
technologies that provide relatively coarse, general information about
a large area to site specific technologies that provide very detailed
information about increasingly smaller areas of the seafloor. These
technologies lend themselves to a "telescope" approach of
investigations with progressive finer detail as needs dictate. This is
the approach generally being followed with the exploration of the EEZ.

6.   WIDESWATH SIDESCAN SONAR SURVEYS

The initial approach to implement the program is to survey the EEZ
utilizing the GLORIA (Geologic Long Range Inclined Asdic) survey system
available through a cooperative agreement with the British Institute of
Oceanographic Sciences. This tool was developed specifically to map
the morphology and texture of seafloor features in the deep ocean,
i.e. greater than 200 m, and has the capability of surveying nearly
27,000 km²/day.

7.   MULTIBEAM TECHNOLOGY

The historical surveying approach limited ocean mapping of large areas
since it could not be performed economically with existing technology.
During the 1960's development was begun on new technological approaches
to ocean mapping. The most significant improvement was the development
of the multibeam or swath mapping technology in the 1960's by the Naval
Oceanographic Office and the General Instrument Corporation. Multibeam
mapping systems are capable of projecting multiple individual beams of
sound laterally in a fanshaped array from the ship through a number of
transducers. The resultant beams, when processed aboard the survey
ship, produce a "swath" coverage of the seafloor. The width of this
swath is a function of the water depth.
     To provide the necessary high degree of positioning accuracy
required for detailed mapping with the multibeam mapping systems,
shore-based medium range electronic systems such as ARGO and Raydist
are used. A significant improvement in positioning capability will
occur when the Global Positioning System (GPS) becomes operational with
24 hour coverage. The fully operational GPS constellation will greatly
expand the capabilities to survey the outer continental shelf,
continental slope, and the remote areas of the EEZ off Alaska and the
Pacific Trust Territories.

8.   CONCLUSIONS

This paper has primarily concentrated on the near term ocean mapping
requirements to meet the urgent, priority needs of users. It is based
upon using technology available to NOAA and the USGS. However, it is

recognized that the vastness of the EEZ region, combined with the
limitations of the current technology, will require a sustained
continuing effort, including the introduction of new technology, in
order to properly characterize and map the seafloor of the EEZ.

While it is difficult to forecast future events, there are
indications that the seafloor will be increasingly utilized by the
Nation for a variety of civilian and military purposes. Long-range
leasing plans from the MMS indicate that the deeper waters of the
continental shelf and slope, i.e. water depths of 2,000 m, will be
developed with increased frequency. Alaska, Pacific island
territories, and other frontier regions will continue to be explored
for energy resources.

In the hard mineral arena, polymetallic sulfides, manganese and
cobalt enriched crusts and nodules, and near shore placer deposits will
be considered as sources of strategic materials. Among potential
mineral-rich regions in the EEZ are the Blanco Fracture Zone, and Gorda
Ridge regions off the coast of Oregon and Washington, the cobalt
enriched crusts on the flanks of seamounts off the Hawaiian Archipelago
and Blake Plateau off the east coast.

Other uses such as waste disposal, seafloor habitats, ocean
thermal energy, bio-technology, deep water pipelines, and cable routes
will need high quality seafloor maps. Bathymetric maps may
significantly improve the efficiency of bottom fishing.

The full potential of the ocean can be achieved when the
characterization phase is completed. Seafloor maps, like topographic
maps on land, are essential for the development of ocean resources.
They guide the identification, assessment, and eventual production of
resource on or beneath the seabed. The national program of systematic
mapping of the ocean floor has been underway by NOAA and USGS since
1984. The program is continuing and will develop a base of information
and maps that will be used for the development of ocean space. New
technology is being evaluated and will be integrated into the program
as it becomes operational.

9.   SOURCE MATERIAL FOR FURTHER READING

Bossler, J.D. and Lockwood, M. (1986) 'The Exclusive Economic
    Zone: A kaleidoscope of issues facing ocean development',
    Coastal Zone '85 Conf. Proc., July 1985.

Compass Publications (1987) 'Special Report - Ocean engineering
    and resource development in the Exclusive Economic Zone', Sea
    Technology 28(6), June, pp.77.

EEZ-Scan 84 Scientific Staff (1986) 'Atlas of the Exclusive
    Economic Zone, Western Conterminous United States', US
    Geological Survey Miscellaneous Investigations Series I-1792,
    pp.152, Scale 1:500,000.

EEZ-Scan 85 Scientific Staff (1987) 'Atlas of the Exclusive
  Economic Zone, Gulf of Mexico and Easter Caribbean Area`,
  US Geological Survey Miscellaneous Investigations Series
  I-1864-A, pp.104, Scale 1:500,000.

Hill, G.W. and Lockwood, M. (1987) 'Seafloor exploration and
  characterization: Prerequisite to ocean space utilization`,
  Oceans 87 Conf. Proc., Marine Technology Society, Washington
  D.C., pp.724-729.

Hill, G.W. and McGregor, B.A. (1988) 'Small-scale mapping of the
  Exclusive Economic Zone using wide-swath side-scan sonar`,
  Marine Geodesy, Vol. 12, pp.41-53.

Laughton, A.S. (1981) 'The first decade of GLORIA`, J. of
  Geophysical Research, Vol. 86, pp.11511-11534.

Marine Board (1987) 'Technology requirements for assessment and
  development of hard mineral resources in the US EEZ`, Working
  Paper, Committee on Technology Requirements for the Exclusive
  Economic Zone Utilization, National Academy of Engineering,
  National Academy Press, Washington D.C., pp.179.

Marine Technology Society (1986) 'Sonar technology for science
  and commerce`, (7 Individual Papers), MTS Journal Special
  Issue, Vol. 20(4), Washington D.C., pp.72.

McGregor, B.A. and Offield, T. (1983) 'The Exclusive Economic
  Zone: An exciting new frontier`, US Geological Survey,
  Washington D.C., pp.20.

McGregor, B.A. and Lockwood, M. (1985) 'Mapping and research in
  the Exclusive Economic Zone`, jointly published by the US
  Geological Survey and the National Oceanic and Atmospheric
  Administration, Washington D.C., pp.40.

National Advisory Committee on Oceans and Atmosphere (1986) 'The
  need for a national plan of scientific exploration for the
  Exclusive Economic Zone`, Final Panel Report, NOAA, Washington
  D.C., pp.62.

National Oceanic and Atmospheric Administration (1986)
  'Exploring the new ocean frontier`, The Exclusive Economic Zone
  Symp. Proc., NOAA, NOS, Charting and Geodetic Services,
  Rockville, MD, Oct. 1985, pp.270.

Rowland, R.W., Gould, M. and McGregor, B. (1983), 'The US
  Exclusive Economic Zone - A summary of its geology, exploration,
  and resource potential`, US Geological Survey, Circular 912,
  pp.24.

US Congress, Office of Technology Assessment (1987) 'Marine
    minerals: exploring our new ocean frontier', OTA-0-342,
    Washington D.C., pp.347.

US Department of the Interior (1984) 'A national program for the
    assessment and development of the mineral resources of the
    United States Exclusive Economic Zone', Symp. Proc., USGS
    Circular 929, Washington D.C., pp.308.

US Department of the Interior and US Department of Commerce (1988)
    'Planning for the next 10 years', US Exclusive Economic Zone
    Symp. Proc., USGS Circular 1018, Reston, Virginia, pp.179.

University of Hawaii/National Science Foundation (1987) 'Ocean
    engineering technology in the Exclusive Economic Zone', results
    of a workshop on "Engineering Solutions for the Utilization of
    the Exclusive Economic Zone Resources", Univ. of Hawaii, Oct.
    1986, pp.18.

Vogt, P.R. and Tucholke, B.E. (1986) 'Imaging the ocean floor:
    history and state of the art', in P.R. Vogt and B.E. Tucholke
    (eds.), The Geology of North America, Vol. M, The Western North
    Atlantic Region, Geological Society of America. pp.1-27.

THE POTENTIAL RESOURCES OF THE SEA AREAS AROUND THE REMAINING
DEPENDENCIES OF THE UNITED KINGDOM

PROFESSOR ALEC J. SMITH
Department of Geology
RHBNC (University of London)
Egham Hill, Surrey
U.K.

1.   INTRODUCTION

The United Kingdom government has not proclaimed zones of economic
exclusivity (EEZs) for the seas contiguous to the United Kingdom, nor
for any of the United Kingdom's remaining dependencies, arguing that
the already declared exclusive fishing zones and, as a matter of
international law, the inherent rights to the continental shelves make
such a declaration unnecessary.
     The potential EEZs of the United Kingdom and its dependencies
total nearly two million square nautical miles and include regions in
the Atlantic, Indian, Antarctic and Pacific Oceans (Figure 1 and Table
1) with examples therein of every sea-floor geological condition.   The
purpose of this paper is to consider what resources, particularly non-
living resources may exist in the unproclaimed EEZs of the United
Kingdom's dependencies.

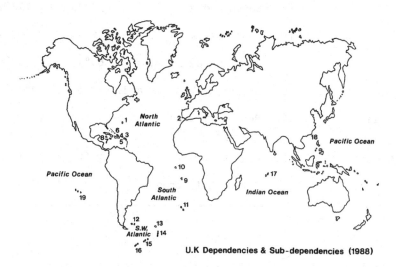

U.K Dependencies & Sub-dependencies (1988)

Figure 1   (See Table 1 for names and latitudes and longitudes).

D. A. Ardus and M. A. Champ (eds.), Ocean Resources, Vol. I, 21–48.
© 1990 Kluwer Academic Publishers. Printed in the Netherlands.

TABLE 1. UK Dependencies and Sub-Dependencies

			Latitude/Longitude (Approximate)	
North Atlantic	1.	Bermuda	32°19′N	64°44′W
	2.	Gibraltar	36°09′N	5°21′W
Caribbean	3.	Anguilla	18°12′N	63°05′W
	4.	British Virgin Isles	18°30′N	64°20′W
	5.	Montserrat	16°45′N	62°10′W
	6.	Turks & Caicos Islands	21°45′N	71°50′W
	7.	Cayman Islands	19°20′N	81°15′W
(Central America)	8.	Belize	17°30′N	88°40′W
South Atlantic	9.	St. Helena	15°57′S	5°42′W
	10.	Ascension	7°59′S	14°24′W
	11.	Tristan da Cunha	37°05′S	12°17′W
Southwestern Atlantic	12.	Falkland Islands	51°45′S	60°00′W
	13.	South Georgia	54°30′S	37°00′W
	14.	South Sandwich Islands	57°45′S	26°30′W
	15.	South Orkey Islands	60°40′S	45°30′W
	16.	South Shetland Islands	62°00′S	60°00′W
Indian Ocean	17.	British Indian Ocean Territories	7°00′S	72°00′E
Pacific Ocean	18.	Hong Kong	22°24′S	114°15′E
	19.	Pitcairn Islands	25°00′S	130°06′W

TABLE 2.   Harvested Living Resources

(Source 1984 Yearbook of Fishery Statistics Vol.58 F.A.O. Rome 451pp) -
Nominal Catches in Tonnes x 1000.

North Atlantic	Bermuda	8.00 (1978) declining to 0.5 (1984). Mainly tuna.
	Gibraltar	Negligible
Caribbean	Anguilla	Negligible
	British Virgin Islands	0.3 remaining steady 1979 to 1984. Mainly Fish and Crustacea
	Montserrat	0.1 remaining steady 1979 to 1984. Mainly Fish.
	Turks & Caicos Islands	1.0 remaining steady 1979 to 1984. Mainly lobster and conches.
	Cayman Islands	0.5 remaining steady 1979-1984. Various.
(Central America)	Belize	1.5 (approx) remaining steady 1975-1984. Lobster, conches and fin-fish.
South Atlantic	St.Helena	0.2 remaining steady 1979 to 1984. Mainly Tuna and Mackerel.
	Ascension	No statistics.
	Tristan da Cunha	No statistics.
South Western Atlantic	Falkland Islands	) No landings recorded
		) by Japanese and Soviet
		) Union fleets together
	South Georgia	) with catches by
		) fleets from Korea,
	South Sandwich Islands	) Poland, Germany (DR)
		) Germany (FR), Chile
	South Orkney Islands	) Argentina,Bulgaria and
		) Spain
	South Shetland Islands	)
Indian Ocean	British Indian Ocean Territories	No statistics
Pitcairn Islands	Hong Kong	150.8 (1975) rising to 199.6(1984). Various of all types.
	Pitcairn Islands	Negligible.

TABLE 3. Possible occurrences of non-living resources
(after Smith & Doherty, 1988)

REGION	DEPENDENCY	AGGRE-GATES	PLACERS	PHOSPH-ORITES	Mn NODULES	Co CRUSTS	POLY-METALLIC CRUSTS & SEDIMENTS	HYDRO-CARBONS
North Atlantic Ocean	Bermuda	*	–	*	*	–	–	–
	Gibraltar							
Caribbean	Anguilla) Brit. Virgins)	*	*	*	*	–	–	–
	Cayman	*	–	–	*	–	*	–
	Montserrat	–	*	*	–	–	*	–
	Turks and Caicos	*	–	*	–	–	–	–
	Belize	*	*	*	–	–	–	*
South Atlantic Ocean	Ascension	*	–	*	–	–	*	–
	St.Helena	*	–	*	*	*	*	–
	Tristan da Cunha	–	–	–	*	–	*	–
Antarctic	Falklands	*	–	*	*	–	–	*
	Sth.Georgia	–	–	–	*	–	*	–
	Sth.Sandwich	–	–	–	*	–	*	–
	Sth.Shetlands	*	*	*	–	–	*	*
	Sth.Orkneys	–	–	*	*	–	*	–
Indian Ocean	Br.Indian Ocean Territories	*	–	*	*	–	–	*
Pacific Ocean	Hong Kong	*	–	–	–	–	–	*
	Pitcairns	–	–	–	*	*	–	–

In contrast to the detailed knowledge of the zone contiguous to the United Kingdom, the zones around the dependencies are barely explored and their resource potential can only be assessed by extrapolation from better known regions. It is suggested that a statement of interest in the potential EEZs of the dependencies will stimulate research and, perhaps, a detailed evaluation by interested parties. In what follows each dependency, or group of dependencies, will be considered in turn and a best estimate of resources will be presented.

2.   THE UNITED KINGDOM (the zone contiguous to Great Britain)

Before discussing the possible resources of the various dependencies, it is appropriate to comment on UK contiguous zone. The extent of any area that may be proclaimed as an EEZ is limited by the proper claims of adjacent nations across the North Sea, English Channel and Irish Sea. The northwestward extension into the Atlantic Ocean is further complicated by conflicting claims to the uninhabited islet of Rockall (Lat.57°36′N, Long.13°36′W). Sufficient to say here that the contiguous continental shelf has already yielded living and non-living resources of very considerable wealth to the United Kingdom, principally through living resource and hydrocarbon and aggregate exploitation. By the efforts of such government departments and agencies such as the Ministry of Defence, the Ministry of Agriculture, Fisheries & Food, and the British Geological Survey, various laboratories, the academic community and industrial companies, the contiguous shelf and seas and seafloor around the United Kingdom and out to beyond Rockall are amongst the best surveyed in the world.
     In addition to a complete suite of hydrographic charts, prepared by the Hydrographer to the Royal Navy, geological maps at a scale of 1:250,000 showing superficial sediments, Quaternary geology and solid geology together with gravity and magnetic data have been prepared by the Marine Earth Sciences and Hydrocarbons Division of the British Geological Survey. The deep geology of this region has also been thoroughly investigated with data, in places, revealing details to depths of 70 km. Studies related to the exploration for, and exploitation of, hydrocarbons together with other studies have led to thorough understanding and evaluation of the geology of the UK continental shelf and adjacent continental slope. Furthermore surveys by the British Geological Survey and commercial companies concerned with aggregate recovery have provided details of workable seabed sand and gravel reserves. The recovery of hydrocarbons and aggregates make a major contribution to the economy of the United Kingdom.
     The exploitation of the living resources from the seas around Britain has a long history and the scientific investigation of these resources has long been important. The UK fishing industry has been unsettled in recent decades due to the loss of distant water fishing grounds but currently the inshore fishing industry is more buoyant than it has been. More catches are being landed and sales directly to foreign factory ships are a notable feature. Fish farming is being expanded rapidly, particularly around the northern shores.

3.   THE DEPENDENCIES AND SUB-DEPENDENCIES OF THE UNITED KINGDOM

The remaining British dependencies are the last remnants of a once
global empire; they are territories which remain linked to the Mother
country for a variety of reasons ranging from political expediency to
the continuation of a status quo because no other pathway presents
itself. Many of them are remote - both geographically and in
circumstance - from their parent country. Most have been neglected
except when external forces have raised issues of national pride or
political expediency. The wealthier, in relative terms, dependencies
of the World War II era have passed into independence often assuming
Commonwealth status and several of these, particularly in the southern
Pacific, are already benefitting from the 1982 Law of the Sea
Convention.
     In marked contrast to the knowledge gained of the seas around the
United Kingdom, the seas around the dependencies are poorly catalogued.
Local detailed knowledge abounds but systematic surveys have yet to be
conducted. Because of their global distribution and varied
climatological, oceanographic and geologic settings, the seas around
the dependencies offer exciting challenges. What follows concentrates
on the non-living reserves, however Table 2 shows, in general terms,
the near current state of the fishing industries of the dependencies.
Only Hong Kong, soon to become part of China, has a well developed
fishing industry backed by extensive research. The Falkland Islands in
the south western Atlantic Ocean are known to be in the centre of
extensive fishing grounds though until very recently, those islands
gained little from that source. Instead, distant fishing fleets from
many countries, especially from Japan and the Soviet Union, had taken
advantage of those grounds. Now a British distant water fishing fleet
is beginning to work there and all countries must pay for licences to
fish within the 150 mile fishing zone.
     So far as the non-living resources are concerned, Table 3 attempts
to show the range of possible deposits. It is, because of the lack of
evidence, vague. More details are given when describing the
dependencies later is this paper. Much of the discussion is based on a
report (Smith and Docherty, 1988) prepared for the Marine Technology
Directorate Limited of the United Kingdom, whose generosity in
permitting the use of that report is hereby acknowledged.

3.1 Dependencies of the North Atlantic

Two dependencies in very different settings are located in the North
Atlantic region. They are Bermuda, which is of volcanic origin and
rises from abyssal depths, and Gibraltar, which is a small prominent
limestone peninsula off southern Spain.

Bermuda   Latitude 32°19′N, Longitude 64°44′W (Fig. 2)
          Population (1980): 67,700

The Bermudan islands and islets lie near the southeast margin of a
truncated late-Tertiary volcanic cone which rises from the floor of the
western North Atlantic about 600 nautical miles south east of New York.

The summit of the extinct volcano is about 80 m below present sea-level
and forms a platform of about 500 km² extent. At the times of
Quatenary low sea-levels, calcareous dunes were created from exposed
coral reefs and these now-cemented dunes are the main cause of the
island's elevation above sea-level (Moore & Moore, 1946). The present
day corals form a veneer over and around fixed dunes. Once off the
edge of the platform, depths increase rapidly to about 4 km. From that
isobath the ocean floor slopes gently out to a depth of 5 km - the

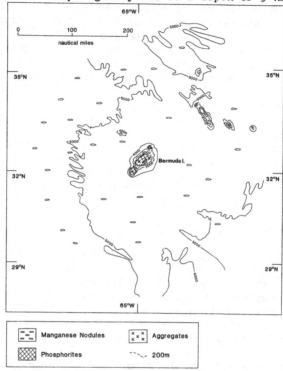

Figure 2.   Bermuda: bathymetry (in metres) and potential resources.

whole feature making the Bermuda Rise which lies between the Sohm,
Hatteras and Nares abyssal plains to the northeast, west and south
respectively. The relatively flat ocean floor between the 4 km and
5 km isobath is interrupted by occasional abyssal hills which shoal to
depths of 2 km.
    The fishing industry based on Bermuda has been in decline with
tuna catches falling dramatically in the period 1978 to 1984
(F.A.O.1986) presumably because of the general pattern of over-fishing
in the North West Atlantic region. The potential for improvement in
this industry might be considerable but is outside the scope of this
paper.
    The seas around the islands already have an enviable reputation
for sport-fishing: this and other recreational assets must be
preserved.

The mineral wealth of any possible EEZ, which could be more than 125,000 square nautical miles (nm²) in extent*, appears to be limited. Carbonate sand could, with care, be dredged in places and there may be guano-derived phosphate crusts in pre-Holocene deposits. In deeper waters there is the likelihood of manganese rich crusts and nodules - the area being sufficiently remote from the detrital input from the North American continent. Little detailed work has been done in the vicinity of the islands but the area merits further surveys and the proclamation of an EEZ may be a necessary stimulant.

Ocean Thermal Energy Conversion (OTEC) offers a future source of power for the main island, the temperature difference between surface and ocean bottom water being sufficient to provide power. The high capital cost of an OTEC plant, when set against the current price of conventional fuels, makes early installation unlikely.

<u>Gibraltar</u> Latitude 36°09′N, 5°21′W
           Population (1982): 32,000

Gibraltar is a narrow peninsula extending from the Iberian mainland into the Alboran Sea on the east side of the Bay of Algeciras. Because of the proximity of neighbouring countries and conflicting claims for what would be a relatively small portion of the western Mediterranean, no attempt is made here to discuss existing and potential resources.

3.2 The Caribbean Dependencies

These comprise of a widely scattered suite of islands. Anguilla, the British Virgin Islands and Montserrat are located at the northern end of the Lesser Antilles archipelago, the Turks and Caicos Islands lie towards the southeastern end of the Bahaman archipelago and the Cayman Islands rise from the Yucatan Basin (Case, 1975).

The islands enjoy broadly similar climatic regimes though their geological and oceanographic settings are different.

<u>Anguilla</u> Latitude 18°15′N, Longitude 62°54′W (Fig. 3)
          Population (1970): 6,500

The main island, some 70 km², and associated islets - Dog, Seal, Scrub, Little Scrub, Prickly Pear Cays and, some 20 miles to the north west, Sombrero - are located on the drowned Anguilla Bank and are composed of Lower Miocene reefal limestones resting on Eocene-Oligocene volcanic rocks which form part of the active Lesser Antilles volcanic arc. Anguilla would share with the British Virgin Islands a common EEZ which would be restricted to the west, south and east by other legitimate claims but extend northwards from the Lesser Antilles volcanic arc, across the eastward extension of the Puerto Rico trench, out on to the Nares abyssal plain. The size of such an EEZ may not exceed 25,000 nm².

* All EEZ areas are only very general estimates: much careful consideration will be required to determine the precise area and extent of the zones.

Anguilla has purely local fishing interests and sport fishing for
tourism is poorly developed. Further developments should be possible
both on the Anguilla Bank and beyond but access to markets is limited.
The island has a long established salt industry using natural
evaporation of sea water in salt pans. The local salt could be used to
develop a salted fish industry.

The potential for offshore minerals include dredged carbonate
sands from the Bank, the possibility of some buried placer deposits
and, beyond the trench, manganese nodules on the red clay of the Nares
abyssal plain. There may be metalliferous seeps on the inner slope of
the trench, though these may prove to be more of a scientific curiosity
rather than an economically viable deposit.

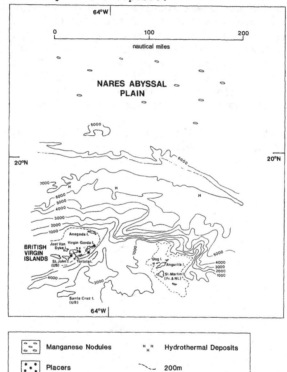

Figure 3. British Virgin Islands and Anguilla: bathymetry and
potential resources. (Fr = French; NL = Netherlands; US = USA)

British Virgin Islands   Latitude 18°30′N, Longitude 64°20′W (Fig. 3)
                          Population (1980): 12,000+

These islands consist of the most easterly of the Virgin Islands group.
Those to the extreme west are administered by Puerto Rico, the central
group by the USA. The BVI comprise Tortola, Virgin Garda, Anegada,
together with the smaller islands of Ginger, Cooper, Salt, Peter,
Norman, Tobago, Jost Van Dyke, Guana and Great Camanoe. Their total
area is 100 km². The islands are part of a volcanic arc created

between Jurassic and Neogene times. The rocks of the islands are pre-Albian Cretaceous volcanics, Upper Cretaceous sandstones, volcanics and limestones, Eocene andesites and derived sandstones with limestones, Oligocene-Miocene mudstones and reefal limestones. The whole is a product of a complex evolution with an early subduction phase followed by strike-slip faulting. The Anguillan islands are part of the same geological story.

A small fishing industry is located in the British Virgin Islands, mainly for fish, of several species, and crustacea. This could be expanded with more investment. Tourism, too, is important in these islands and sport fishing is already important.

So far as non-living resources are concerned, like Anguilla there are prospects for manganese nodules in the Nares abyssal plain and metalliferous seeps on the slope of the Puerto Rico trench. Of further interest, there could be placer deposits since in the past copper, associated with molybdenum, has been worked on Virgin Gorda from an intrusive complex. A similar complex occurs on Tortola. At times of lower sea-level placer concentrations may have occurred.

Sands and gravels of various compositions - siliciclastic, volcanic and carbonate - are dredged for local use already.

Montserrat   Latitude 16°45′N, Longitude 62°10′W (Fig. 4)
              Population (1980): 12,000++

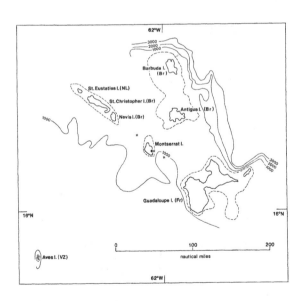

Figure 4. Montserrat: bathymetry and potential resources.
(Br = British Commonwealth; Fr - French; VZ = Venezuelian)

Montserrat is located towards the inner, western, edge of the Lesser
Antilles Ridge. The island, some 100 km² in extent, is volcanic with
the highest of seven volcanic centres, reaching 1000 m above sea-level.
Evidence shows episodic volcanism since Late-Miocene times with a range
of volcanic rocks, mainly andesites and basalts, displayed. The
present island represents the part of a once larger and subsequently
eroded complex (MacGregor, 1938).
     Fishing is an essentially local industry and not very significant;
it seems possible that it could be further developed. Non-living
resources include offshore sands and the possibility of hydrothermal
mineralisation and placer deposits related to the history of volcanic
activity and the effects of sea-level change in the Quaternary. There
may also be guano-derived phosphate deposits on the shelf.
     The size of an EEZ will be restricted to about 2000 nm² because of
the proper claims of neighbouring territories including that of
Venezuela based on the ownership of Aves Island on the Aves Ridge.

Cayman Islands   Latitude 19°15′N, 81°15′W (Fig. 5)
                 Population (1979): 16,700

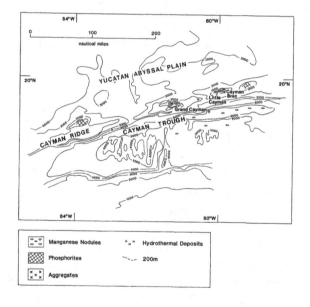

Figure 5.   Cayman Islands

Situated in the Caribbean Sea between Cuba and Jamaica, the Caymans
consists of three islands, Grand Cayman and, approximately 100 km to
the east-north-east, Little Cayman and Cayman Brac. The total land
area is about 250 km². The islands, composed of recent to Oligocene
limestones mainly coral or coral derived sands, are located on the
prominent Cayman Ridge, a relatively narrow elongated escarpment-like
feature with a depth, except where above sea-level, of around 1-2 km.
To the north is the Yucatan abyssal plain with depths exceeding 4 km

while immediately to the south of the ridge, and aligned parallel to
it, is the Cayman Trough with depths exceeding 5 km before shoaling to
1 km on the Nicuragua-Jamaica Rise.

The EEZ rights of neighbouring islands in the Greater Antilles
limit the size of the EEZ which could be proclaimed for the Cayman
Islands to about 40,000 nm².

Fishing is moderately important already, but could be improved
considerably. Upwelling water should give a high biological
productivity hereabouts. Tourism is already important and can be
further improved.

Small quantities of carbonate sand are dredged and supplies are
adequate for local demand. High organic productivity may have given
rise to phosphatic accumulations while in deeper waters manganese
nodules may occur. The Cayman ridge and trough are the site of left-
lateral movements at a plate margin: in such a circumstance
polymetallic sulphides from hydrothermal mineralisation become targets
for future exploration.

Turks and Caicos Island   Approximately between Latitude 21° and 22°N
                           and Longitude 71° and 72°30′W (Fig. 6)
                           Population (1980): 7,500

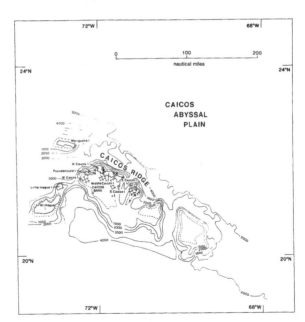

Figure 6. Turks and Caicos Islands

These islands are located on the Caicos Ridge at the south-eastern end
of the Great Bahaman Bank and consist of the four main Caicos Islands -
Providinciales, North, Middle and East Caicos and the two smaller
islands of West and South Caicos, strung from northwest to southeast -
and the much smaller Turks Islands lying to the south east of the Turks

Island Passage. The islands have a total area of about 430 km² and are composed of modern reefal carbonates resting on carbonate bedrock ranging in age from Cretaceous to Pleistocene. Fishing, particularly of lobster, is important and could be developed further. Non-living resources include phosphates which are thought to have been derived from guano. Some of these phosphate deposits occur in caves. Dredged carbonate sands are another resource but use is purely local. Once off the ridge the water deepens rapidly into the Hispaniola trough, at the western end of the Puerto Rico trench, to the south and the Caicos abyssal plain, part of the Nares abyssal plain, to the north. Some hydrothermal deposits may occur in the slopes of the Ridge and there may be manganese nodules in the abyssal plain to the north.

The prospective EEZ is limited to the south by the proper claims of Haiti and the Dominican Republics but the total area may exceed 25,000 nm².

3.3   Central America

The United Kingdom retains one crown colony, Belize, on the mainland of the Americas. Its sovereignty is disputed by Guatemala and there have been numerous attempts to settle the problem by peaceful negotiation and military force. The issue remains unresolved.

<u>Belize</u> (British Honduras)   Latitude 17°30′N, Longitude 88°40′W
                                   (and around)
                                   Population (1980): 140,000+

Belize has an area of 22,700 km² (approx.), is situated on the east coast of the Central American isthmus facing the Caribbean Sea and bordering Mexico to the north and Guatemala to the west and south. The northern region is geologically part of the Yucatan peninsula; the southern region is mountainous rising to 1,120 m in the Maya mountains.

The immediate offshore area is dominated by a barrier reef near the shelf break some 15 to 30 nautical miles offshore. The reef, a major hinderance to shipping, hosts a prolific marine life. Beyond the break the seafloor deepens rapidly into the Gulf of Honduras, the south western extension of the Cayman Trench, itself related to one of the major geological features of the Caribbean (see Cayman Islands).

The possible EEZ is hard to define because of conflicting claims by Mexico, Guatemala and Honduras and may not exceed 10,000 nm². There is already a well developed local fishing industry and this could be developed further. The possible mineral resources could include hydrocarbons in the relatively narrow shelf and continental slope.

3.4   The South Atlantic

Great Britain has three dependencies in the South Atlantic, all widely separated but all in broadly similar settings, that is, they rise as ancient volcanoes from the ocean floor between transform fracture zones on the flanks of the Mid-Atlantic Ridge (Baker, 1973; Emery and Uchupi, 1984).

St. Helena   Latitude 15°57'S, Longitude 5°42'W (Fig. 7)
             Population (1982): 5,500

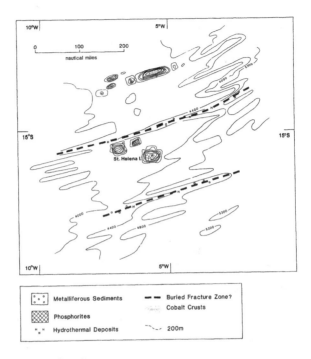

Figure 7. St. Helena Island

The island is the eroded summit of a large composite volcano of Early
Tertiary age composed of basaltic tuffs and lavas cut by trachytic and
phonolitic intrusions (Hirst, 1951). Surrounded by steep cliffs, the
highest point on the island is about 900 m above sea level and its area
is 128 km². The island lies about 800 nautical miles east of the South
Atlantic ridge crest and between the St. Helena and Martin Vaz
transform fracture zones.
      Fishing has been a feature of the island but catches are low and
consist mainly of tuna and mackerel. Opportunities for development
exist. Manganese ore exists on the island, it is of limited extent,
and there are small occurrences of iron ore and wind blown shell sands.
Offshore there may be calcareous and volcanic sands to be dredged. In
deeper waters there is little likelihood of manganese nodules to the
west since the ocean floor depth may not meet depositional
requirements; to the east prospects may be better. Metalliferous
sediments and hydrothermal deposits, related to the volcano's early
history, may exist as might cobalt crusts on some sea-mounts to the
north of St. Helena. The two fracture zones may also have hydrothermal
sulphides associated with them.
      The total EEZ which could be claimed is more than 125,000 nm².

Ascension   Latitude 7°59′S, Longitude 14°24′W (Fig. 8)
            Population (1982): 1625

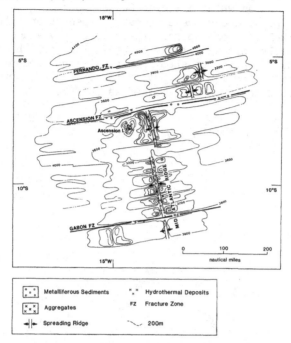

Figure 8.   Ascension Island

Ascension like St. Helena, which lies about 700 nautical miles to the
south east, is an extinct volcanic island (Daly, 1925) associated with
the seafloor spreading evolution of the South Atlantic. It is closer
to the axis of the ridge (under 100 miles to the west of the crest) and
slightly younger in geological age. It has a land area of just under
100 km² and its highest point is 950 m above sea-level. It is composed
of basalt and alkaline trachytes. Though no coral reefs surround the
island, many of the beaches are composed of near pure calcareous sands.
It might be added that the island has a major airport, mainly for
military purposes, and it is an important staging point on the air
route to the South Western Atlantic Ocean.
      The living reserves around the island are untapped and there are
no statistical records of catches. The non-living resources of the
narrow shelf could include calcareous sands and the occurrence of
offshore phosphates: there are small phosphate deposits, derived from
guano, on the island.
      The most significant features within the possible EEZ are the
crest of the Mid-Atlantic Ridge and the Ascension Fracture Zone, one of
the transform faults which cross the ridge to the north of the island.
A second, unnamed and smaller, fracture zone lies to the south of the
island. Mineral deposits are known from the Mid-Atlantic Ridge in the
TAG (26°N) and FAMOUS (37°N) areas. There, hydrothermally deposited
manganese-oxide crusts and concretions, with sediments enriched in iron

and copper, occur in vents and immediately adjacent zones. The ridge
here, however, is spreading at a slow rate as compared with some of the
world's richer ridges which are also faster spreading. Further,
because the ridge is above the carbonate compensation dilution of the
deposits is likely. There do not appear to be any sea-mounts in the
area, consequently only the flanks of Ascension could have cobalt
crusts. Because of detritus shed from the island, economically
workable crusts are unlikely to have formed. Again a total area of
more than 125,000 nm² could be claimed as an EEZ.

<u>Tristan da Cunha</u> Latitude 37°05'S, Longitude 12°17'W (Fig. 9)
                Population (1982): 325

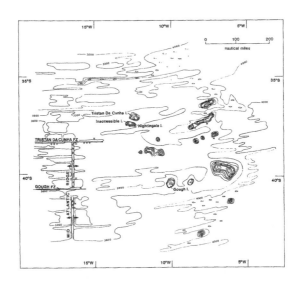

Figure 9. Tristan da Canha and associated islands. (For symbols see
previous figures)

Tristan da Cunha is one of a group of islands in the South Atlantic
east of the Mid-Atlantic Ridge. Only Tristan is populated, the others,
Inaccessible and Nightingale just south of Tristan and Gough nearly 200
nautical miles to the south east, are rarely visited. Tristan, some
120 km in area, is formed of a single primary stratovolcano rising to a
height of over 2,000 m above sea-level and composed of lava flows and
pyroclastics which range in rock type from andesite to basalt, with
trachybasalt predominating. The last eruption was in 1961.
Inaccessible and Nightingale (and the latter's associated islets of
Middle and Stollenhoff) are similar in origin and rock type to Tristan.
All were formed as part of the spreading ocean ridge system and lie at
the southwestern end of the Walvis line of sea-mounts which stretches
from the African coast and are linked to 'hot-spot' activity. Gough
Island is similar in composition to the others. Like the others it has

steep cliffs and it rises rapidly to over 1,000 m. There are no
minerals of economic importance there (Le Maitre, 1958).
    There are no statistics relating to fishing - the seas around the
islands are severe and winds frequently of gale force.
    Within an EEZ, which because of the dispersed nature of the
islands could exceed 250,000 nm^2 there is a portion of the Mid=Atlantic
Ridge, a number of sea-mounts and two fracture zones - the Tristan da
Cunha and Gough. Mineralisation associated with these features is
possible, the sea-mounts are unlikely to host cobalt crusts because of
the high dissolved oxygen content of the waters, and manganese nodules
will be restricted to the deeper (>4 km) areas of the seafloor. The
ocean floor morphology hereabouts is not known in significant detail.

3.5 The Southwestern Atlantic and Antarctic Islands

The Falkland Islands, South Georgia, South Sandwich, South Orkney and
South Shetland Islands lie in the southernmost Atlantic Ocean with the
more southerly bordering Antartica. Their origins are various, as will
be discussed below, and their ownership is contested.

The Falkland Islands   Latitude between 51°00 and 52°30′S
                       Longitude between 57°30′ and 61°30′W (Fig. 10)
                       Population (1980): 1,900

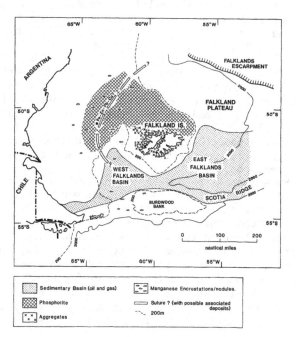

Figure 10. Falkland Islands.

The islands consist of two main islands West Falkland and East Falkland (3,500 and 5,000 km² respectively) separated by the Falkland Sound, and about 200 smaller islands. The islands rise from the Falkland Plateau, an eastwards extension of the Argentine continental shelf. To the north the plateau is bounded by a steep slope, the Falkland escarpment, while to the south the west-east aligned West Falklands and East Falklands Basins separate the plateau from the Burdwood Bank.

The rocks of the Falkland Islands range in age from possibly Precambrian to Early Cretaceous. West Falkland and the northern part of East Falkland consist of lower Palaeozoic sedimentary rocks intruded by basaltic dykes. The remainder consists of Mesozoic sandstones and mudstones. The offshore basins contain a thick, up to 9 km (Greenway, 1972) succession of late Mesozoic and Tertiary sediments.

Conventional opinion states that the Falkland Plateau is a spur of the South american continent; recent discoveries however link the geology of the Falkland Islands with that of southeastern Africa (Mitchell et al, 1986). If the latter view is correct, and there is much to support it, then there is the possibility of a collision suture between the islands and the South American continent.

The living reserves around the Falkland Islands are immense. The waters are rich in marine life and a major fishing industry has been developing rapidly. Few fish are landed at Stanley, the capital and main port of the Islands, but licencing of fishing in the 150 mile fishing zone led in 1987 to £15 million accruing to the islands, and it is reported that squid alone fished by foreign fleets has exceeded £400 million in value in one year. The fishing fleets of the Soviet Union, Japan, Spain, both Germanies, the Eastern Block countries, South Korea and other nations as well as fleets from adjacent Argentina and Chile have all been engaged in catching a wide variety of species. As mentioned earlier, a British distant-water fleet has also commenced fishing in this region. The population of the Falkland Islands has gained immensely but now concern is being expressed (F.A.O. Fisheries Technical Paper 286, 1987) about over-exploitation. Proper licencing backed by effective patrolling is the only solution which could work in everybody's interests, especially if it is backed by appropriate investment in research.

The offshore mineral prospects of the Falkland Islands have already merited exploration by major oil companies which have investigated the oil and gas potential of the offshore sedimentary basins. The fact that the sovereignty of the islands is the continued subject of dispute between the United Kingdom and Argentina inhibits further research but the chances of finding major plays must be regarded as good. It is known that a suitable reservoir sandstone exists in the Cretaceous succession (the Springhill formation) of the nearby Magallanes Basin, and the possibilities of similar reservoirs in the largely unknown overlying Cretaceous and Tertiary successions must be high. More research is needed into source rocks and reservoir structures.

Nearer to the islands there could be shell deposits sufficient to compensate for the lack of limestone formations on the islands while in deeper water phosphorite deposits may be of economic potential. Diatomaceous oozes could be a possible source of nucleation for

phosphate deposits. Manganese crusts and nodules may also occur. The
existence of a suture between the Falkland microplate and the Southern
American continent, if recent thinking is correct (Martin, 1986),
offers the prospect of placer deposits in sediment-filled valleys on
the platform. Much more research is obviously required.

A total EEZ of more than 150,000 nm² could be claimed around the
Falkland Islands, even allowing for the proximity of the South American
mainland.

Falkland Islands Sub-dependencies

South Georgia   Latitude between 54° and 55°,
                Longitude between 36° and 38°W (Fig. 11)
                Population (1980): 22 together with visiting fishery
                                   workers

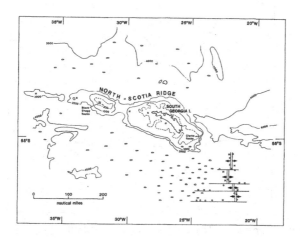

Figure 11.   South Georgia Island and associated rocks.

Some 170 km by 30 km in extent, its highest point reaches just above
3,000 m. Geologically it is part of a late Jurassic to early
Cretaceous island-arc/back-arc basis system correlated with the
Southern Andes. It is made up of ophiolites, basement and andesitic
derived sediments with gabbroic and granitic intrusions. It lies on
the North Scotia Ridge which is bounded to the north by the eastward
extension of the Falklands Trough.

As with the Falkland Islands, the seas around South Georgia abound
with marine life with the abundance of krill being an important part of
the food chain. Whaling was for a long time a major industry around
the island. Prospects for further development of the fishing industry
are good and, no doubt, major catchesare recovered from the seas
hereabouts.

No mining activities have taken place in the island, but, because
of its plate-boundary setting, recent, as well as geologically older,
mineralisation with polymetallic sulphides could be significant.
Further offshore, manganese nodule and crust formation could be
favoured because of oxidising bottom currents.

The total size of any prospective EEZ may reach 250,000 nm².

South Sandwich Islands   Latitude between 56°00´ and 59°30´S, between
                         Longitude 26°00´ and 28°15´W (Fig. 12)
                         Population: nil

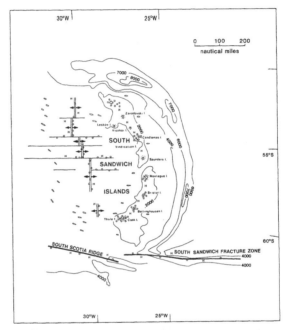

Figure 12.   South Sandwich Islands.   (See previous figures for symbols)

The South Sandwich Islands comprise a volcanic arc of eleven islands -
from north to south - Zavodovski, Leskor, Visokoi, Candlemas,
Vindication, Saunders, Montagu, Bristol, Bellinghausen, Cook and Thule.
Their total area just exceeds 700 km². Cook, Bristol, Montagu and
Visokoi are ice-covered; Saunders and Candlemas are more than half ice
covered and the remainder are free of ice-caps. Most of the islands
show recent or continuing volcanic activity and there are submerged
volcanic cones at the northern end of the arc. Basaltic and andesitic
rocks predominate, some of the islands are simple cones, the others are
more complex (Baker, 1978). There is an area of active spreading
´behind`, i.e. west of, the chain and a deep arcuate trench, the South
Sandwich trench, to the east. To the south there is the main South
Scotia Ridge and South Sandwich fracture zone. These features, the
trench and the fracture zones, and the North Scotia Ridge, which
extends eastwards from South Georgia, form parts of the margin of the
Scotia Plate.

The marine living resources of the South Sandwich arc are very considerable and though climatic conditions are inhospitable, vessels of a number of countries already harvest this marine life.

Geological conditions promise a variety of resource possibilities: hydrothermal metalliferous deposits from the back arc spreading ridge and associated small transform faults; polymetallic sulphides may occur on the flanks of the Scotia Ridge and in the vicinity of the volcanic arc. Manganese-rich oxides with a range of other metallic elements can occur on the deep ocean floor and cobalt crusts could occur on the flanks of the islands and sea-mounts.

Volcanic arcs are known to contain metallic ore bodies and these may, therefore, exist on the narrow shelves of the islands. Research may reveal the presence of a porphyry copper deposit.

Mining on land at these southern latitudes would be extremely difficult; marine mining, even if a rich ore body should be found, may verge on the edge of the impossible.

The possible extent of an EEZ could exceed 180,000 nm².

3.6   British Antarctic Territories

The South Orkney Islands and South Shetland Islands are part of the British Antarctic Territory which became a colony separate from the Falkland Islands and its sub-dependencies in 1962. The Territories also include the Antarctic Peninsula (Graham Land) and a portion of Antarctica. Much has yet to be decided about future sovereignty but the South Orkney Islands and South Shetland Islands are discussed here for the sake of completeness. The possible size of an EEZ could reach 250,000 nm².

South Orkney Islands (Fig. 13) comprise a group of ice covered islands around latitude 60°40′S and between longitudes 44°30′and 47°00′W. They lie on the southern limb of the Scotia Ridge with the closest being some 300 nautical miles north east of the Antarctic Peninsula. There are four larger islands - Signey, Coronation (the largest), Laurie and Powell - and several smaller islands. Geologically, they consist of a metamorphic basement of pre-Carboniferous age succeeded by Carboniferous to Cretaceous sedimentary rocks. Sulphide mineralisation has been reported from Coronation island in the metamorphic rocks and later dykes (Matthews & Malling, 1967).

There is abundant marine life around the islands but prevailing conditions make fishing and whaling hazardous. Offshore mineral deposits, including hydrothermal veining and polymetallic crusts, are possible but must be regarded as unworkable.

Much the same can be said about the South Shetland Islands (Fig. 14) which stretch between latitude 61°00′ and 63°00′S and longitude 63° to 54°W. The total area of the islands is about 4,200 km²; from southwest to northeast they are named Smith, Low, Deception, Snow, Livingston, Greenwich, Robert, Nelson, King George, O′Brien, Aspland, Gibbs, Elephant and Clarence with King George and Livingston being the largest. They lie northwest of the Antarctic Peninsula being separated by the Bransfield Strait with its active

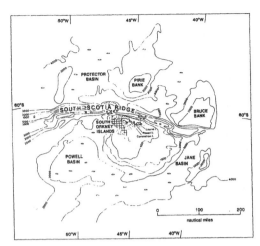

Figure 13.   South Orkney Islands.

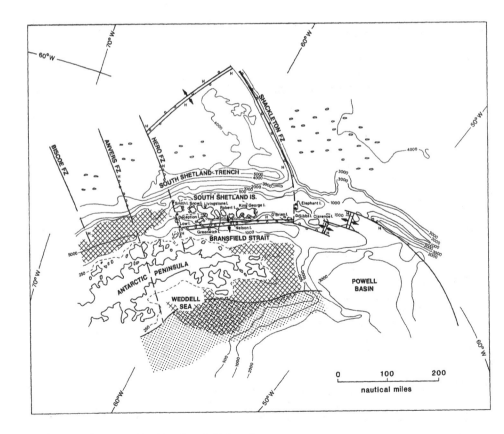

Figure 14.   South Shetland Islands.

back-arc spreading centre and there is a series of cross-cutting
transform fracture zones (Holdgate & Baker, 1979), Northwest of the
islands is the South Shetland trench on which subduction has apparently
ceased. (Smellie et al, 1984). Further oceanwards lies the Scotia Sea
with its own spreading centre. The most easterly islands are formed of
metamorphic rocks: the others, except Deception, which is an active
volcano, are formed largely of igneous rocks and volcanically derived
sediments. As might be expected in such a geologically active region,
mineralisation is extensive. It includes porphyry copper deposits,
polymetallic occurrences of copper, lead, zinc and silver and chrome in
a disseminated form. Nickel, cobalt and asbestos as well as tin
sulphides also occur. The entire suite occurring in a series of belts
aligned with the axis of the archipelago (Hawkes, 1982; Rowley &
W. Pride, 1982; Behrendt, 1983).

Living resources offshore are abundant but conditions are
hazardous; non-living resources will reflect those known from on land.
The latter, whilst of great interest, are said not to amount to minable
prospects. The offshore deposits may have some degree of marine
concentration but such are the extreme climatic conditions that further
consideration is not merited at this time.

3.7 Indian Ocean

British Indian Ocean Territory (Chagos Group)

Latitude 7°00′S, Longitude 72°00′E (Fig. 15)
Population: no statistics (old established population of more
            than 2,000 removed to Seychelles and Mauritius before
            1975; military personnel only on Diego Garcia).

The archipelago is made up of a group of small islands on the Great
Chagos Bank (a feature about 750 km² in extent) - Nelson, Three
Brothers, the Eagles, Danger and Egment - and several other islands to
the north (Peros Banhos, Solomon Islands) and one (Diego Garcia) to the
south. They lie at the southern end of the Laccadive - Chagos Ridge
which is taken to represent a foundered micro-continent of the ancient
super-continent of 'Gondwanaland' (Heezen & Thorp, 1965).

The islands are composed of coral and detrital carbonate sands
resting on volcanic rock, probably basalt (Francis & Shor, 1966), the
whole bank is probably subsiding and a number of previous islands have
already drowned (Laughton et al, 1970). From a depth of around 200 m,
the sea floor falls eastwards to the abyssal plain of the Central
Indian Basin, at depth exceeding 5 km, whilst to the west the floor
falls to about 4 km and then begins to rise again on the eastern flank
of the actively spreading Carlsberg Ridge, the crest of which lies
about 300 nautical miles from the islands.

There is a wide range of marine living resources in what is an
upwelling water region, but little is currently know to be exploited.
So far as non-living resources of a potential EEZ are concerned, the
Central Indian Basin, which is at the outer limits of any EEZ has

already yielded ore-grade samples of manganese nodules with the nickel, copper and cobalt content reaching 2.47[5] (Frazer & Wilson, 1980). Other non-biological resources are phosphate deposits related to the upwelling and phosphorites are known to occur further north.
The possible size of an EEZ could exceed 200,000 nm².

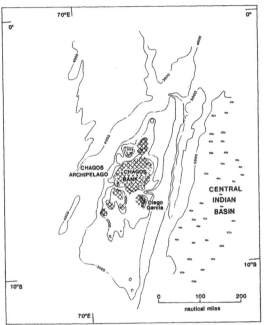

Figure 15. British Indian Ocean Territories

3.8 Pacific Ocean

In this vast region only two colonies remain - Hong Kong, which will soon be assumed by the Peoples' Republic of China, and the Pitcairn group of island in the remote southern Pacific.

Hong Kong Latitude 22°24′N, Longitude 114°15′E
           Population: >4,000,000

Situated on the Kwangtung coast of S.E. China, this 1,032 km² territory reverts to China in 1997, consequently no detailed discussion is presented in this account.
The fishing industry has been a key part of the economy for decades and fishing boats from Hong Kong work both the nearby shelf and much more distant waters. This industry is well developed and has been backed by much research.
Non-living resources offshore include sand, gravel and aggregates. Hydrocarbon prospects of the shelf have been investigated in the South China Sea but so far the results are disappointing.

<u>Pitcairn Islands</u>   Latitude 25°00'S, Longitude 130°06'W (Fig. 16)
                 Population (1983):   61

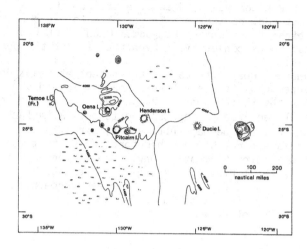

Figure 16.   Pitcairn Island and associated islands.

Pitcairn Island, detailed above, is the only populated island of a
group of remote islands in the southeastern Pacific Ocean. Pitcairn,
5.2 km² in area and rising to 333 m, is entirely volcanic in origin
being composed of alkali basalts and minor trachytes some 0.45 to 0.93
million years in age (Duncan et al, 1974). It is the youngest volcanic
member of the Tuamotu Chain and may be near a 'hot spot'. (Recently
submarine volcanic activity has been observed south-east of Pitcairn
thus supporting the idea of a northwestwards movement of the earth's
lithosphere hereabouts.) The other islands, Oneo, Ducie and Henderson
(Latitude 24°20'S, Longitude 128°20'W) are corraline, with Ducie, the
most remote being about 250 nautical miles to the east. The possible
EEZ approaches 200,000 nm².
        The living resources of this part of the Pacific Ocean are largely
undocumented but Japanese, South Korean and fishing vessels of other
countries do range into this part of the Pacific.
        Within any potential EEZ there are banks and sea-mounts as well as
abyssal depths away from the chain. Cobalt crusts could be a possible
target though the best prospects are known to be north of Latitude 15°S
and geologically older sea-mounts are considered better prospects than
extremely younger ones. Manganese nodules in the abyssal plains are
another possible target so long as the ocean floor is deeper than the
carbonate compensation depth (CCD) at which all carbonate debris is
taken into solution. Here the CCD is at about 4,500 m depth.

4.   CONCLUSIONS

The purpose of this paper was to draw attention to the possible rewards
which would follow from declaring EEZs for what is the political entity
of the Dependencies of the United Kingdom. Much of what has been
described has had to be conjective. Even the size of the possible
zones is speculative.
However, where a real interest in the potential resources of the
seas around a dependency has been taken - namely around the Falkland
Islands following the Shackleton Report of 1982 - very considerable
benefits have followed. Declaring EEZs would be a stimulant to further
research, evaluation, and perhaps ultimately, exploitation. Since many
of the dependencies are remote and therefore remote from markets,
economic forces will dictate most strongly the timing of any form of
exploitation but without first declaring the interest, little progress
can be expected. What is clear is that an area approaching
2,000,000 nm² could be claimed and much valuable scientific, as well as
economic, information collected.

5.   BIBLIOGRAPHY

Baker, P.E. (1973) 'Islands of the South Atlantic', in
    A.E.M. Nairn and F.G. Stehli (eds.), The Oceans Basins and
    Margins, 1: The South Atlantic, Plenum Press, New York and
    London, pp.493-553.

Baker, P.E. (1978) 'The South Sandwich Islands: III', Petrology
    of the Volcanic Rocks', Sci. Rept. Br. Antarct. Surv., No. 93.

Behrendt, J.C. (1983) 'Petroleum and mineral resources of
    Antarctica', USGS Circular 909.

Case, J.E. (1975) 'Geophysical studies in the Caribbean Sea', in
    A.E.M. Nairn and F.G. Stehli (eds.), The Oceans Basins and
    Margins, 3: The Gulf of Mexico and the Caribbean, Plenum Press,
    New York and London, pp.107-180.

Csirke, J. (1987) 'The Patagonian fishing resources and the
    offshore fisheries in the South-West Atlantic', FOA Fisheries
    Technical Paper, 286, Rome, pp.75.

Dalry, R.A. (1925) 'The geology of Ascension Island', Proc. Amer.
    Acad. Arts and Sci., 60, No. 1, pp.1-80.

Emery, K.O. and Uchupti, E. (1984) 'The Geology of the Atlantic
    Ocean', Pringer Verlag, New York.

FAO (1986) 'Yearbook of fishery statistics: 1984', 58, Rome,
    pp.451.

Francis, T.J.G. and Shor, G.G. Jr. (1966) 'Seismic refraction measurements in the north-west Indian Ocean`, J. Geophys. Res., 71, pp.427-449.

Frazer, J.Z. and Wilson, L.L. (1980) 'Manganese nodule resources in the Indian Ocean`, Marine Mining, 2, No. 3, pp.257-292.

Greenway, M.E. (1972) 'The geology of the Falkland Islands`, Sci. Rep. Br. Antarct. Surv., No. 76.

Hawkes, D.D. (1982) 'Nature and distribution of metalliferous mineralization in the northern Atlantic Peninsula`, J. Geol. Soc. London, 139, pp.803-809.

Heezen, B.C. and Tharp, M. (1965) 'Physiographic diagram of the Indian Ocean` (with descriptive sheet), Geol. Soc. Amer. Inc., New York.

Hirst, T. (1951) 'Observations on the geology and mineral resources of St. Helena`, Overseas Geology and Min. Res. 2, pp.116-176.

Holdgate, M.W. and Baker, P. (1979) 'The South Shetland Islands: I. General description`, Sci. Rep. Br. Antarct. Surv., No. 91.

Laughton, A.S., Mathews, D.H. and Fisher, R.L. (1970) 'The structure of the Indian Ocean`, in A.E. Maxwell (ed.), The Sea, 4, New York: Wiley-Interscience, pp.543-586.

Le Maitre, R.W. (1958-59) 'The geology of Gough Island, South Atlantic`, Overseas Geology and Min. Res., 7, pp.371-380.

MacGregor, A.G. (1938) 'The volcanic history and petrology of Montserrat, with observations on Mt. Pele, in Martinique`, The Royal Society Expedition to Montserrat, BWI, Phil. Trans. R. Soc. London Ser. B, No. 557, 229, pp.1-90.

Martin, A.K. (1986) 'Microplates in Antarctica`, Nature, London, 319, pp.100-101.

Matthews, D.H. and Maling, D.H. (1967) 'The Geology of the South Orkney Islands: I. Signy Island`, Sci. Rep. Falkland Island Depend. Surv., No. 25.

Mitchell, C., Taylor, G.K. and Shaw, J. (1986) 'Are the Falkland Islands a rotated microplate?`, Nature, London, 319, pp.131-134.

Moore, H.B. and Moore, D.M. (1946) 'Preglacial history of Bermuda`, Bull. Geol. Soc. Am., 57, pp.207-222.

Rowley, P.D. and Pride, D.E. (1982) 'Metallic mineral resources of the Antarctic Peninsula (Review)', in D. Craddock (ed.), Antarctic Geoscience, Univ. Wisconsin Press, Madison, pp.859-870.

Smellie, J.L., Pankhurst, R.J., Thompson, M.R. and Davies, R.E.S. (1984) 'The geology of the South Shetland Islands: VI. Stratigraphy, geochemistry and evolution', Sci. Rep. Br. Antarct. Surv., No. 87.

Smith, A.J. (1986) 'Opportunities and management of EEZs: A view of the UK position' in Exclusive Economic Zones, Advances in Underwater Technology, Ocean Science and Offshore Engineering, 8, Soc. Underwater Technology, Graham & Trotman, London, pp.99-128.

Smith, A.J. and Docherty, J.I.C. (1988) 'Non-living resources of potential EEZs of British Dependencies', Report for Marine Technology Directorate Limited, London, pp.49.

INVESTIGATING FRANCE'S EEZ: MAPPING AND TECHNOLOGY

GUY PAUTOT
IFREMER
Centre de Brest
BP 70, 29263 Plouzane
France

A mapping program involving mapping of sea-bed resources must answer a certain number of preliminary questions: the answers have an effect on one another.

- what is the surface area to be mapped?

- what is the range of depths?

- what accuracy is required?

- what is the time limit for completing the project?

    Replies to these questions are prerequisites for determining the most efficient means for data acquisition. The purpose here is to look into this subject by bringing up the technological needs as seen through the user's eyes.

    In the program under consideration the area to be mapped is extensive: 11,256,000 km², that is to say 20 times that of metropolitan France. This area is the sum of 16 non-contiguous regions spread over the three great oceans. Round the mother country, the Exclusive Economic Zone (EEZ) represents 340,000 km². The French EEZ off the Americas in the Atlantic ocean represents 356,000 km². The French EEZ in the Indian ocean covers 2,885,000 km² and that in the Pacific ocean stretches over 7,675,000 km².

    Apart from the surface area, we must examine the vertical scale for the ocean depths. We have drawn up a hypsometric chart of the French EEZ with three units. Defined as lying between 0 and 300 m, the continental shelf covers about 900,000 km², i.e. 8% of the total surface. Defined at the zone between depths of 300 and 3000 m, the margin unit represents about 3,600,000 km², i.e. 32% of the total surface. Set between average depths of 3,000 and 5,000 m the ocean deeps represent about 6,750,000 km², i.e. 60% of the whole surface area.

    A first conclusion is that the ocean deeps represent the main part of the surface to be mapped, and that the continental shelf section is much smaller. This has implications for the strategy to be followed

D. A. Ardus and M. A. Champ (eds.), Ocean Resources, Vol. I, 49–56.

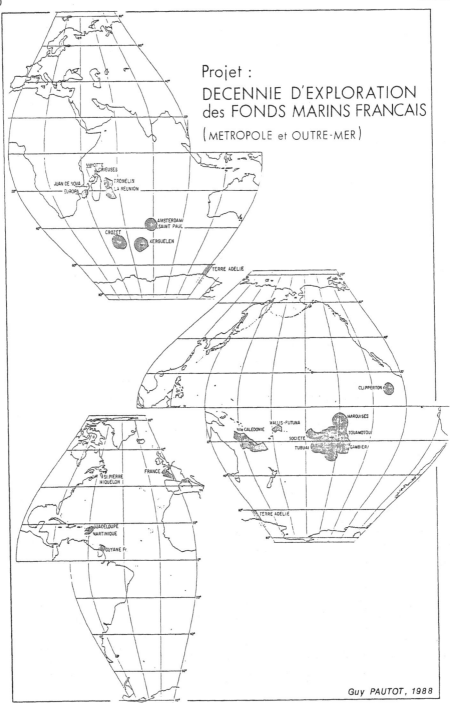

Projet :
DECENNIE D'EXPLORATION
des FONDS MARINS FRANCAIS
(METROPOLE et OUTRE-MER)

Guy PAUTOT, 1988

Figure 1.   Project: Decennie d'exploration des fonds marins Francais
(Metropole et Outre-Mer).

and the technology to be used. On the other hand the margin unit
includes the slope, which is a rather unknown area, difficult to map
and which will require a suitable methodology. In this analysis we
are interested mainly in the water depths deeper than 1,000 m: ocean
deeps and lower part of the margins.

A period of ten years is expected to be needed to carry out the
first program. Why ten years, and is it feasible with our available
means? One rough guide to the answer is that with the Gloria side
looking sonar, which is able to survey 25,000 km^2 per day, we could
survey the 300 to 5,000 m deep zone (i.e. 92% of the total area) within
about 420 days. Although this may be feasible as part of a complete
and systematic survey, it would still only provide one sort of
information. Then again, working the mineral deposits of the ocean
deeps, if this is to be considered within the next one or two decades,
would mean beginning the field work program now.

The objectives in view must also be considered with mapping
representation in mind, and especially with attention to the
appropriate scales to be agreed upon.

This scale selection requires a few explanatory remarks. Existing
data (morphology, geology) are very disparate and navigation accuracy
of the 1970-1980 ocean campaigns, carried out by help of the satellite
TRANSIT System, is roughly 500 m. This level of precision involves
uncertainties in plotting the vessel position of 5 mm on a map at a
scale of 1:100,000, of 2 mm for a scale of 1:250,000 and of 0.5 mm at
1:1,000,000. Choice of spacing of isobaths also depends upon the
selected scale. We generally use a 1,000 m or 500 m spacing on a map
drawn at a scale of 1 to 1 million and a 250 m to 100 m spacing on a
map drawn at a scale of 1:250,000. For instance, using the Seabeam
multibeam echo-sounding system greatly improves the determination of
depth distribution, but graphic representation does not allow more than
three isobaths per millimeter to be drawn without prejudice to
legibility.

In the French program the practical choice is to proceed with
scales of 1 to 250,000 around France, and 1 to 1 million for the
Overseas territories and Provinces. Isobaths will be drawn at
intervals of 100 or 50 m for a scale of 1 to 250,000, and with a 500
and 200 m spacing for a scale of 1 to 1 million. Representation may be
performed according to more accurate graduated data (Seabeam type) but
while keeping the richness of the sea bed morphological
characteristics.

The horizontal resolution aimed at is about 100 m (possible with
GPS) and the vertical resolution of the order of 50 m.

This choice implies that detailed studies, carried out with an
accuracy appropriate to scales from 1:10,000 up to 1:100,000 may be
integrated into the maps by degraded representation, but this is not a
first phase objective of the exploration program. In any case, when
industrial applications become relevant, further detailed studies of
specific targets will have to be carried out.

The large area to be mapped, the varied range of depths to be
investigated, the expected periods of time involved, and the level of
resolution required for the bathymetric (and morphological) data lead
to the definition of an appropriate methodology and choice of mapping

instruments. Satellite data are suitable for defining the great
morphological units within data poor areas. The useful scale for
representing this type of data cannot be more detailed than 1 to 5
million. For this program it will thus be necessary to use a vessel
whose survey speed is between 8 and 12 knots: the purpose is to draw
bathymetric, morphological, geological maps, illustrate the character
of the sea bed, and identify areas of natural risks or induced risks.

To reach these targets within the prescribed delay, on the one
hand it will be necessary to use a technology which enables the various
parameters to be mapped together at high speed and, on the other hand,
to give priority to "surface" surveys over "linear" surveys in order to
reduce as much as possible the need for extrapolations, and to cover a
maximum area per unit of time.

This concept of attaching as much importance as possible to
efficient use of vessel time is a basic point for our program, and has
implications for the technology needed.

Choice of mapping equipment in this paper will be made in
accordance with the user's objectives and not according to criteria
defined by the engineer in charge of the technological side. By using
equipment already available and by developing their performances we
believe that our proposals are not unreasonable.

Regarding the bathymorphology of the ocean deeps (with exception
of the continental shelf) let us compare four available systems:
Seabeam, Hydrosweep, Seamarc 2 and Gloria. These operate differently
and yield different products.

Gloria can survey about 22,000 km² per day at 10 knots. It is the
equipment enabling the most extensive coverage within a given time.
Deficiencies involve restriction of the data to imagery only, absence
of bathymetry, presence of a blind lane under the vessel that must be
filled up through interpolation by computer, the length of the shadow
cast on the external section of the beam which suppresses structures, a
useful resolution which does not generally justify maps more detailed
that 1 to 500,000.

The average scanning range of the Seamarc 2 sonar is 5 km on each
side. This enables surveys of about 4,500 km² per day to be made,
i.e. a surface five times smaller per unit time than Gloria covers.
However, it combines acoustic imagery of the sea bed and a bathymetry
of average precision of the order to 50 m. The narrower area swept
means a higher angle of incidence of the external part of the beam and
thus a lesser shadow and better useful resolution.

The Hydrosweep multibeam sounder provides bathymetry to a lane
whose width is equal to twice the water depth, that is, the lane width
is, for example, 8 km when the depth reaches 4,000 m. In deep water,
the range is thus similar to that of the Seamarc 2 sonar, with a
scanned surface of about 3,500 km² per day. Whereas Seamarc 2 is
towed, the Hydrosweep array is hull mounted, allowing easier operation
and higher speeds (up to 15 knots). Depth accuracy varies from 10 to
20 m. It does not provide any acoustic imagery of the sea bed.

The Seabeam is also hull mounted. Its vertical accuracy is about
10 to 20 m. It "scans" a narrower width than Hydrosweep (equal to
about 0.8 times the water depth). In a day, at a depth of 4,000 m, and

at 10 knots, it is thus possible to map an area of 1,300 km², again
without seafloor imagery.

With the EEZ program in view, we are in need of equipment to
obtain acoustic imagery of the sea bed and bathymetry simultaneously,
and with the widest possible lateral coverage and highest operating
speeds (10-12 knots) in order to survey the maximum surface area per
unit time.

The best equipment would combine properties of Gloria and
Seamarc 2. Like Seamarc 2, it should enable both a seafloor image and
bathymetric data to be collected. Bathymetry should be accurate to
about 20 to 50 m so as to be able to exploit fully a scale of 1 to
250,000. The axial shadow area of the sonar should be "lit" by a
multibeam acoustic array. The area swept on each side would be less
than the 30 km it is for Gloria, so as not to overstretch the shadows.
A Seamarc 2 type equipment with a 10 km range on each side would enable
about 9,000 km² to be surveyed per day: this would place it within the
efficiency gap that needs to be filled.

The other technological element decisive for the program concerns
the determination of sea bed characteristics and the assessment of
mineral deposits.

Various parameters have an influence on sea bed reflectivity with
respect to insonification by sonar systems: microtopography, sea bed
roughness, type of terrain. It is obvious that separating the
influence of each of these three parameters, added to that of the
shadow effect of structures that can mask large areas, is not a simple
problem. First, it requires visual exploitation of the data by an
engineer who is a specialist in interpreting raw information, and
correlating this with so called "ground truth". Exploitation may be
helped by suitable processing to change relevant parameters into
pseudo-colors. However the actual systems still have limitations.
There is still no efficient way, for instance, to identify manganese
nodules and other polymetalic deposits on the sea bed acoustically.

Various types of echo-sounders are used for mapping nodule fields.
The 3.5 kHz sounders determine the thickness of the transparent
superficial level which may, in some instances, be correlated with the
presence of nodules. In addition, new multifrequency acoustic arrays
have been built and used: such as the Sumitomo 100 A - MFES system and
the Raytheon MBSP system. In principle, these multifrequency sounders
enable the nodule diameter and their abundance on a vertical line right
under the moving vessel to be determined. Tests of these methods by
direct sampling have given unequal results according to the areas
studied.

A multifrequency acoustic array has also been used on the Blake
Shelf to determine the presence and thickness of the oxide and
phosphate deposits. Results of a first experiment seem to be
promising.

Even if this type of array may be able to analyse and quantify sea
bed reflection ratios and, through careful calibration, achieve fine
differentiations, a linear vertical projection can be obtained only
under the best conditions.

As for sea bed acoustic imagery and bathymetry, extension of
existing techniques needs to be explored for analysing the physical and

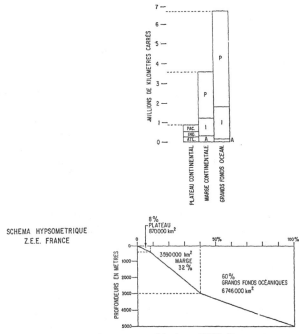

Figure 2.   Histogramme des superficies ZEE France.

SURVEY COVERAGE km² / day (at 10 knots and 4000 m water depth)					
IMAGERY			4 400	~10 000	22 000
BATHYMETRY	1 300	3 500	4 400	~10 000	
TOOLS	Sea Beam	Hydrosweep	Seamarc-2	★ Non-existent to be developed	Gloria

Figure 3.

geological characteristics of the sea floor terrain, as well as the
growth of mineral deposits, and this with arrays covering large
surfaces per unit time. Some tests have been performed on Seabeam data
in an attempt to infer the signature of manganese nodule fields.
Reflectivity maps of different areas were produced using the intensity
of the specular return from each ping. The patchiness of the nodule
coverage is shown by definite highs and lows in the reflectivity
pattern. Our experience indicates that this type of analysis is
effective when the sea bed is smooth, and sedimentation is homogeneous.

The equipment to be developed for mineral exploration would be a
multi narrow beam, multifrequency system. Recent improvements in
signal processing will be very helpful; however, we shall not be able
to avoid "calibration" with "ground truth" data.

To conclude, the overriding technological priority for achievement
of the French EEZ program for ocean deeps is a single or coupled system
providing imagery and bathymetry. This system should have the
possibility of being used at high speed (10 knots), have a range of
about 10 km on each side, possess multifrequencies for analysis of the
sea bed characteristics and not present a blind area under the ship.
It should be able to map 10,000 km^2 per day, which would enable the
mapping of almost the whole territory of the ocean deeps belonging to
the French EEZ within ten years by surveying two months per year.

IFREMER owns a Seabeam multibeam echosounding system and a high
resolution sonar towed near the sea bed, called SAR. These two systems
are not adequate for the planned program, and hence feasibility studies
are in progress to determine the characteristics of the most suitable
equipment.

REFERENCES

de Moustier, C. (1985) 'Inference of manganese nodule coverage
    from Sea Beam acoustic backscattering data`, Vol. 50, No. 6,
    pp.989-1001.

EEZ-Scan 84 Scientific Staff (1986) 'Atlas of the Exclusive
    Economic Zone, Western Conterminous United States`, US
    Geological Survey Miscellaneous Investigations, Series I-1792,
    Scale 1:500,000, p.152.

Farcy, A., Voisset, J.M., Augustin, P. Arzelies (in press)
    'Acoustic imagery of the sea bed`, Offshore Technology Conf.
    1988.

Liao, Y. (1988) 'Multi-purpose acoustic system aids manganese
    nodule exploration`, Sea Technology, pp.33-36.

Patterson, R.V. (1967) 'Relationships between acoustic back-
    scatter and geological characteristics of the deep ocean floor`,
    J. Acoust. Soc. Am., 46, pp.756-761.

Pautot, G. (1988) 'Décennie d'exploration des fonds marins
    français (métro-pole et Outre-Mer)', Rapp. Int. Ifremer, p.79.

Porta, D. (1983) 'Ferromanganese deposits detected by multi-
    frequency acoustic system', Sea Technology, pp.15-20.

Renard, V. and Allenou, J.P. (1979) 'Seabeam, multi-beam echo-
    sounding in Jan Charcot. Description, evaluation and first
    results', Int. Hydrogr. Rev., LVI(1), pp.35-67.

Stanton, T.K. (1984) 'Sonar estimates of seafloor microroughness',
    J. Acoust. Soc. Am., 75-3, pp.809-818.

Sumitomo Metal Mining Co. Ltd. 'Multi-frequency exploration
    system (MFES)', Tech. Inf. 24-8, 4-Chome Shimbashi, Minato-ku,
    Tokyo, Japan.

DEVELOPMENTS OF THE GLORIA SYSTEM FOR MORE EFFECTIVE EEZ RECONNAISSANCE

ANDREW GRIFFITHS
Marconi Underwater Systems Limited
Watford
England
U.K.

ABSTRACT. Until recently, there has been little scope for the detailed exploration of the blue water areas of the ocean. In the late 1960's and early 1970's, significant advances in electronics hardware and signal processing methods led to the advent of compact, easily transportable survey systems of widely different application, e.g. seismic reflection profiling, magnetic and gravity surveys and dedicated deep sea side scan sonar vehicles.

The basis of any form of exploration, whether it be for scientific, industrial or defence purposes, is sound knowledge of the terrain to be examined. Now that the techniques exist for large scale mapping of the deep water area, there is an increasing demand for the strategic knowledge of what lies hidden in the ocean depths. Non-renewable resources, particularly hydrocarbons are already being explored for in depths in excess of 1000 m.

Within such a scenario, marine geoscientists have a unique opportunity to map these vast uncharted areas systematically. The approach suggested is similar to that followed by planetary geologists in surveying the Solar System. Namely, to carry out a regional, low resolution survey first over the entire area, and from it to identify sectors that deserve more detailed study using higher resolution systems.

By following this approach, significant features will be found and the resulting map will present the geologist with the essential regional picture prior to detailed investigation of more localised processes. It is difficult to overstress the importance of such regional geological data when one bears in mind that only 2% of the deep ocean has been mapped and yet nearly all the major structures relating to plate tectonic models lie within the deep water zone.

GLORIA is a very long range side-scan sonar system that can provide the marine equivalent of an aerial photograph of the seafloor. What makes it unique is that it can map up to 15,000 km^2 per day, allowing us for the first time to map a significant percentage of the blue water.

D. A. Ardus and M. A. Champ (eds.), Ocean Resources, Vol. I, 57–72.
© 1990 *Kluwer Academic Publishers. Printed in the Netherlands.*

1.   INTRODUCTION

The time is approaching when there will be a significant increase in exploration activity of the continental margin and beyond. As geoscientists we must look to the future and plan the means by which deep-water surveys can be accomplished. To put these blue water areas into perspective, they make up about 71% of the overall surface of the earth, of which only 2-3% has been mapped (Fig. 1).

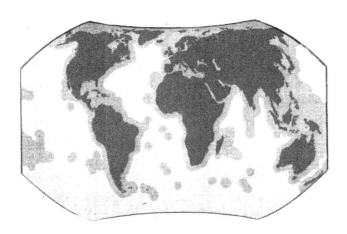

Figure 1. Map of the world with 200-mile limits shown; 36% of sea covered area lies within these limits.

To date, most research carried out has been concerned with studying geological processes and so the localities chosen have been biased to such environments as mid ocean ridges, ocean fracture zones and trenches where subduction is believed to be taking place. In terms of the ocean space this is not a true representation of the greater part of the oceans. In fact there is a danger in "proving" various aspects of such theories as plate tectonics simply because one selects the site for an investigation based on the principles of the theory one is trying to prove. This is not a sound scientific approach and until we have gathered a great deal more data, we cannot be confident that our understanding of the geology of the oceans beyond the continental shelf has any more validity than intelligent guesswork.

Now that we have the capability to study these deep water areas on a large scale using SEASAT and geophysical equipment we must approach marine mapping with a less biased and more evenly spread approach, even though prevailing opinion suggests that areas might have nothing to show. Systematic mapping of the US Exclusive Economic Zone by the USGS using the GLORIA side scan sonar system, has frequently shown that supposedly uninteresting regions have produced some very unexpected results.

The aim of this paper is to consider the best approach to mapping these uncharted oceans by looking at the various types of geophysical equipment available and how that equipment can best be used to provide the most balanced geological picture in the shortest and most cost-effective manner.

2.   THE TECHNIQUES AVAILABLE

By considering first the systems which map on a large scale and then the higher resolution techniques we can follow what is probably the best approach to mapping large ocean areas. The most global mapping system is SEASAT which operated during 1978 and has proved particularly effective in measuring long wavelength gravity anomalies. A number of previously unknown seamounts have been discovered in the Pacific Ocean and the SEASAT data forms part of a first phase map to be used as a base for further more detailed work. It is however limited in its ability to map out more than just broad trends.

By using all the available data a large scale regional survey can then be planned. Although there are several techniques that could be used, the most effective way to provide a regional picture is the long-range side-scan sonar. Of these, the GLORIA system can acoustically map 15,000 km² per day, thus over a typical 30 day cruise, an area of nearly half a million km² can be systematically mapped (Fig. 2).

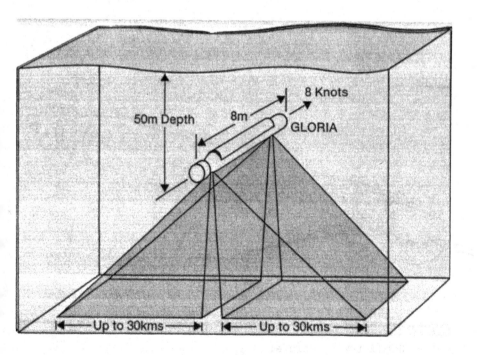

Figure 2.   Schematic diagram illustrating the swath geometry of GLORIA.

3. LONG RANGE MAPPING SYSTEMS

There are a number of different systems which can map the seafloor.
GLORIA is unique in that it operates at a lower frequency than any
other and therefore has a very long range capability. This range
varies depending on the water column characteristics and the depth to
seabed. It is appropriate to describe this system and the type of map
that it produces before considering its value in offshore exploration.

4. GLORIA - GEOLOGICAL LONG RANGE INCLINED ASDIC

The wet end of the GLORIA system comprises a 7.8 m long towfish which
weighs 2 tonnes in air but is neutrally buoyant. It is
hydrodynamically streamlined and designed to give an optimum
performance at tow speeds of between 8-10 knots. On either side of the
towfish there are two rows of 30 transducers (Figs. 3 and 4). These
transmit a 4 second frequency modulated slide over the 100 Hz range
which allows recognition of the returned signal even when it has been
substantially absorbed or dispersed by its interaction with the
seafloor. Above all, it also provides a very much improved range
resolution.

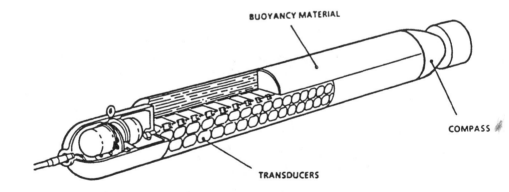

Figure 3. GLORIA towed vehicle.

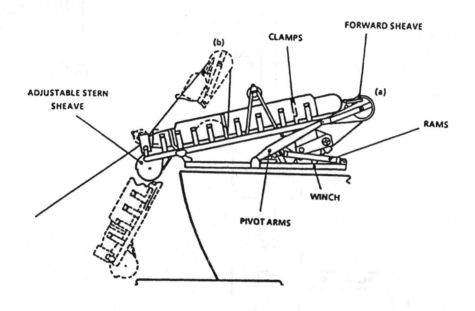

Figure 4. Diagram of vehicle stowage and launching gantry showing principle of operation.

The emitted beam width is 2.5 degrees in the horizontal and 30 degrees in the vertical (at a fixed inclination of 20 degrees from the horizontal). The sampling interval can be varied between 20-40 seconds depending on the range and water depth. The signal emitted into the water has a nominal power output of 10.5 kW on either side. The signal is reflected off the seabed and returns to be picked up by the transducers now used in the receiving mode. An acoustic image of the seafloor (sonograph) is gradually built up by successive scans as the ship moves along the track. The amplitude of the returning pulse depends on the range, type and texture of surficial material and the topography of the seafloor. Any yaw in the vehicle is electronically corrected and because GLORIA is towed at a depth of 50 m the effect of the waves are minimal, except in extreme weather conditions.

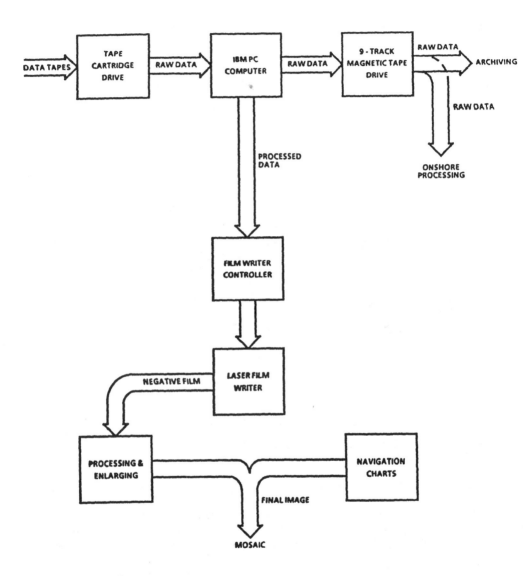

Figure 5.   Onboard image processing system.

5.   DATA RECORDING AND ONBOARD PROCESSING

The data is stored on cassettes and is corrected for radiometric and
geometric distortions on an IBM microcomputer. The processed data is
then written onto film by a specially designed laser film writer. The
resulting photographs are arranged on the navigation chart to produce a
complete mosaic. The raw data is backed up on 9 track magnetic tape
for further processing ashore (Fig. 5). The initial interpretation is
carried out on board by the geologist to ensure quality control and
prevent gaps occurring in the mosaic. The production of an initial
mosaic on board allows for changes in the survey plan when particularly
interesting features are discovered and need to be investigated
further. Figure 6 is a flow diagram illustrating the complete GLORIA
system. The resolution of the Sonograph varies, depending as it does
on water conditions, vessel speed and range, but is roughly 50 m by
150 m. For a more detailed review of the GLORIA hardware, the reader
is referred to Somers et al, 1978.

6.   ONSHORE IMAGE PROCESSING SYSTEM

On completion of the survey, the navigation and GLORIA data stored on 9
track tape are further processed to enhance the image by means of
various contrast stretching techniques. In addition, significant
features identified by the geologist can be enhanced using different
filtering combinations which help to improve the signal to noise ratio.
Colour coding can help to identify any similarities, or variations,
that occur on the sonograph. The geologist uses these and other image
processing techniques to enable him to make a geological interpretation
based primarily on the GLORIA sonograph data, but also utilising the
seismic, bathymetric and magnetic data to produce the best regional map
possible. Once the interpretation has been completed on an individual
area, they are then assembled by the computer to form a complete
mosaic. At this stage adjustments are also made for variations in
signal intensity with range and differences in the direction of shadow
effects due to changes in the track direction. The hardware is based
around the DEC/VAX computer system. Two image processing workstations
are connected to each computer and final print out of the maps is
carried out on a high resolution laser film writer producing large
format negatives with a quality suitable for atlas production. The
image processing techniques software was initially developed from
LANDSAT image processing techniques and has been further adapted to
suit the acoustic underwater environment. A more recent detailed
discussion of developments on image processing techniques has been
published by Chavez P. et al, 1987.

7.   COMPLEMENTARY GEOPHYSICAL SYSTEMS

A sonograph picture of the seafloor can be added to in value when used
with other geophysical equipment. Overlaying a number of other
geophysical data onto the GLORIA mosaic allows the study of the deep

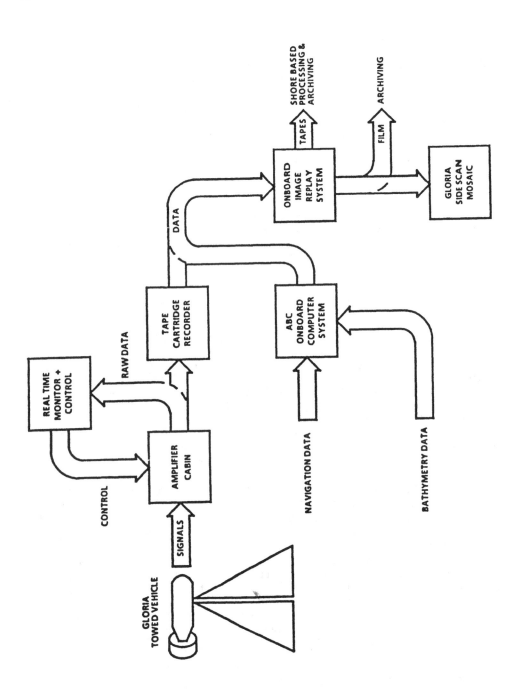

Figure 6.   GLORIA side scan sonar system.

water areas in greater detail than has ever been possible. Seismic, bathymetric, and magnetic data are particularly valuable. By running these systems at the same time and on the same ship as GLORIA, excellent registration may be attained. Typically on board a GLORIA research vessel there is a 12 kHz high precision echo sounder, 3.5 kHz high resolution profiler and a dual channel seismic system. In addition, magnetics and gravity surveys can be carried out simultaneously to complete the picture (see figure 7).

Until recently the benefits of such a programme would have been diminished due to the limited accuracy of the navigation system available in deep water areas. With the development of the Global Positioning System, there is an added incentive to carry out regional surveys in previously unexplored areas.

Having briefly considered the equipment and geophysical techniques involved, let us now consider the potential benefits that reconnaissance mapping can give to a country which has a large open water space.

8.   BENEFITS OF REGIONAL MAPPING

When considering the benefits of an underwater survey, probably the best analogy is that of the use of LANDSAT images when studying the land surface. For example, before almost any activity takes place on the land, a LANDSAT image of the specified area is studied, problems are identified and a plan of action is determined. Similarly, a GLORIA survey provides the base map from which programmes of ocean exploration and development can be planned to be most cost-effective. Prior to high resolution surveys such as multi-channel seismics, swath bathymetry (SEABEAM) or "ground-truthing", a regional GLORIA survey should be carried out to identify those areas of greatest interest. Not only will such a survey save significant time but, in addition, it provides a base map on which separate areas of high resolution data can be overlaid and related. A recent example of how important this can be was during a swath bathymetry survey of the mid Atlantic Ridge in which it proved impossible to interpolate between the track lines even though they were less than 5 miles apart. With a sonograph picture the linear trends could have been followed, but since there was no side scan system a complete contour map could not be produced. In general, all activities relating to the seafloor need the type of information provided by systems such as GLORIA, but it may be useful to illustrate this with some specific examples.

9.   CABLE ROUTE SURVEYS

With the increasing use of fibre optic cables for trans-ocean telecommunication links, the cost of cable laying has escalated considerably. Route planning for such cables becomes important especially in areas that contain geological hazards. A GLORIA survey can help plan the best route and saves days of ship time for the high resolution survey. There are many examples of cables having been laid across active faults, unstable slopes or in the path of fast moving sediments often with catastrophic results. An early Trans Atlantic cable was severed when an earthquake caused a turbidity flow to pass over the cable route. The first cables which crossed the mouth of the Congo broke regularly due to slope instability. A repair ship is on permanent standby in Fiji because the ANZAN cable was laid across an active fracture zone. Today, such risks are unnecessary. For the TAT8, British Telecom commissioned a GLORIA survey off the SW Approaches to map a complex system of submarine canyons prior to laying the cable and help the engineers to identify the best route up the slope.

10. MINERAL EXPLORATION

Before any exploitation of minerals from the deep ocean takes place, a thorough knowledge of their distribution is necessary. To date, for too little study has been made into offshore mineral resource potential and therefore it is very difficult to predict where and how much non-renewable minerals are available. Although it is generally agreed that exploitation is still a decade away, the strategic information of availability and concentration needs to be sought now. The major concentrations of important minerals appear to occur in distinct environments such as on the slopes of seamounts; as polymetallic crusts, polymetallic nodules on the deep sea floor and as sulphide deposits surrounding active hydrothermal vents. All these environments can be easily and rapidly identified from sonograph data using their distinct seafloor characteristics.

Numerous studies of mineral resources, particularly in the Pacific, have been described in the literature (Cronan, D., 1986). A classic study of mineral resource potential by the National Academy of Sciences, USU (Cloud, P., 1969) shows just how dependent society will be on mineral resources from the sea within the next decade.

11. HYDROCARBON EXPLORATION

One of the most immediate uses for GLORIA mosaics is in the preliminary evaluation of potential hydrocarbon environments which have not hitherto been geologically surveyed. As oil exploration moves out to the continental margin environment there is an increasing requirement for more cost-effective exploration techniques. These include utilising computer modelling techniques in sedimentary basin analysis (Parrish, J., 1986). Desk top studies of this kind depend heavily upon valid data being used and very often in new areas this data is simply not available. GLORIA can provide much of this data very quickly and allow much more accurate assessments to be made of hydrocarbon potential. In addition, GLORIA has the ability to provide a major input in the understanding of the regional geology of large areas, which is essential information to any oil company's exploration department beginning work in a new area (Griffiths, A., 1987). East of

Kalimantan, Indonesia there is an area which is opening up to
hydrocarbon exploration. Although the region is in shallow water, it
is surrounded by deep water which has undergone little geological
investigation. There is a plan to map this deep water area by GLORIA,
for an oil company consortium, to help understand the geology of what
is an extremely complex area. The Santos and Compos oil producing
areas in offshore Brazil are known to extend into deeper waters.
Petrobras are already exploring in water depths in excess of 1200 m. A
proposal to map the continental margin in this area to study the
submarine canyons known to exist, would help in understanding the
geological history of the sedimentary sequence on and beyond the
continental margin.

12. WASTE DISPOSAL

The disposal of toxic wastes is an emotive issue, but the fact remains
that there are some hazardous materials that need disposal, and this
should be in the safest possible place. Studies into the suitability
of a number of deep water sites for waste disposal have already been
carried out using GLORIA (Booty, B., 1985), and it appears that stable
deep water areas could be suitable repositories for hazardous waste if
the disposal is carried out properly. However, caution is necessary.
Last year a GLORIA survey beyond the continental shelf of the USA, by
the USGS, mapped a series of active submarine channels which, for the
past decade, had been used as a dumping ground for low level
radioactive waste. Unfortunately, these submarine channels flowed
directly toward a major connurbation and could have been a serious
health hazard simply because the GLORIA survey was not carried out
prior to the dumping of waste to identify a suitable site. Needless to
say, the dumping in that location has ceased.

13. THE REQUIREMENT FOR REGIONAL MAPPING

The need to carry out systematic mapping has been quickly recognised by
the US Government. Under the auspices of the US Geological Survey,
they have initiated a complete GLORIA survey of their Exclusive
Economic Zone. This massive survey has been in progress since 1984
and, to date, has mapped the US Western and Eastern Seaboards, the Gulf
of Mexico, Puerto Rico and parts of Hawaii and Aleutian Island Arc (EEZ
SCAN, 1986 and 1987). The extent of this coverage is shown in figure
8. Once specific areas have been identified by GLORIA as being of
particular interest, the US Government follow up with a programme which
maps small areas with high resolution SEABEAM bathymetry. Although
many countries may not have their resources, there is no reason why
they should not follow the same logical approach on a smaller scale.

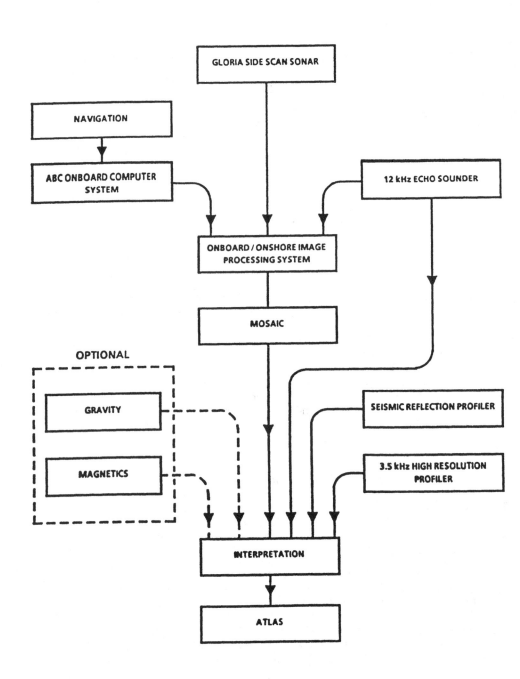

Figure 7. Overall GLORIA system.

13. FUTURE DEVELOPMENTS IN THE GLORIA SYSTEM

Three specific development areas are being actively considered. At present, bathymetry is derived from a precision echo sounder that measures the depth directly below the ship's track. However, since the transducers in the GLORIA towfish are mounted in two sets on either side, it is possible to use the phase difference of the returning signal to the two sets of transducers to calculate the bathymetry along the side scan swath. Although the resolution will be limited by the frequency, a bathymetric contour plot can help in planning for more detailed bathymetric mapping such as that achieved by SEABEAM.

The physical characteristics of the seafloor sediments are becoming increasingly useful to those involved in deep water acoustic propagation studies. It has been long known that an acoustic pulse interacting with the seafloor is distorted. The degree of distortion is a function of the frequency of the acoustic wave, the angle of incidence, the grain size and roughness of the surficial sediments. By comparing the spectrums of the outgoing and returning GLORIA signals, it is possible to deduce the type of seafloor and its acoustic properties. A recent GLORIA survey in the Indian Ocean was able to delineate a manganese nodule field in this way.

The resolution of the GLORIA sonograph is limited by the low frequency of the system. We do not want to change this as it allows us to attain advantages in GLORIA. It is possible however to improve the across track resolution by towing a receive-only array behind the GLORIA towfish. With appropriate channel separation in the hydrophone array and accurate positioning of the two arrays relative to each other, significant improvements can be made to the resolution which may be of benefit to deep sea salvage searches, etc.

The US Geological Survey (Woods Hole) have recently shown the value of being able to incorporate GLORIA and SEABEAM data on the GLORIA image processing facility. By being able to overlay bathymetric, magnetic and gravity data onto the GLORIA base map, there is a significant improvement in the overall regional geological interpretation.

Although these are the developments being introduced at the moment, there is enough flexibility in the system to introduce new technologies as they become available. This is particularly the case with the image processing facility.

14. CONCLUSION

As costs of offshore research escalate and the regions of survey become more hostile and remote, it becomes increasingly necessary to avoid "wild cat" operations and to base survey work on an effective regional database framework. This could be regularly updated. It is now possible to obtain a comprehensive understanding of the processes which form and concentrate resources and thus make exploration much more cost-effective. By carrying out a regional GLORIA survey in the first instance, proper planning which anticipates problem areas can save time and money.

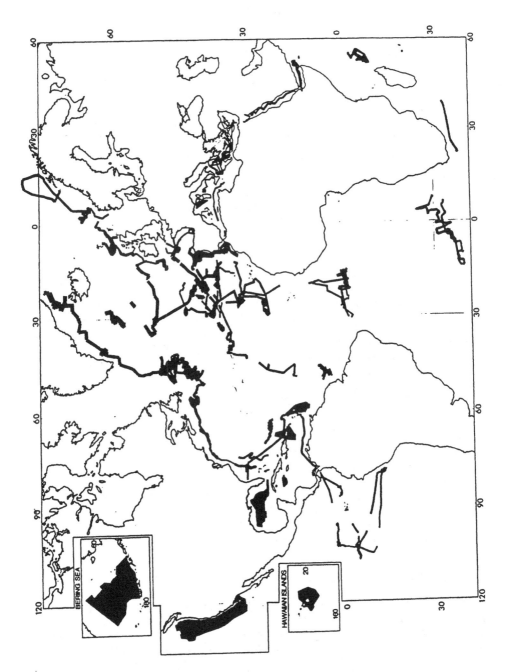

Figure 8. A map of the world showing the GLORIA coverage to date
(Feb. 1987). Note that most of the coverage is in the northern
hemisphere.

By incorporating GLORIA reconnaissance mosaics with high resolution data, a comprehensive knowledge of the deep sea can be attained and a staggering quantity of information from previously uncharted areas becomes available to researchers. In the same way that LANDSAT was a revolution to land based sciences, so GLORIA is to the marine geologists and geophysicists.

GLORIA is unique in its capability of mapping over 15,000 km² per day. By keeping the hardware in the towfish as simple as possible and carrying out the complicated corrections using onboard software, the overall system is extremely reliable, and the equipment has proved itself in extremes of weather conditions. On more than one occasion, GLORIA surveys have continued through hurricanes, although there was a noticeable degradation of data. Another advantage is GLORIA's flexibility. It can be mobilised on any suitable ocean-going vessel in three days. As a system it can be combined with a variety of survey equipment which could be provided by the client. Finally, we would welcome the opportunity of technology transfer in the form of training local geologists and geophysicists into these new techniques. We must take this opportunity to map these unexplored areas in a proper manner. With these types of results we can excite the general public in the last unexplored regions on earth.

15. REFERENCES

Booty, B. (1985) (ed)'Status report into radioactive waste disposal on or beneath the seafloor`, IOS Report No. 204.

Chavez, P., Anderson, J. and Schoonmaker, W. (1987) 'Underwater mapping using GLORIA MIPS`, IEEE Oceans 87, Vol. 3, p.1202.

Cloud, P. (1969) 'Mineral resources from the sea`, National Academy Sci. Report, Ch. 7, San Francisco, p.135.

Cronan, D. (1986) 'Mineral resources in EEZ's`, Adv. in Underwater Tech. Vol. 8, Graham & Trotman, London.

EEZ SCAN 84 and 86 (1986/87) 'Atlases of the US Western Seaboard & Gulf of Mexico`, USGS Misc. Investigations, Scale 1:500,000.

Griffiths, A. (1987) 'A new opportunity for deep water hydrocarbon exploration`, Proc. PacRim Conf. 87, Brisbane, Vol. 1.

O'Driscoll, E. (1981) 'Fundamental structure pattern in disposition of mineral and energy resources`, in Halbourty (ed) Resources of the Pacific Region, AAPG, p.465.

Parrish, J. (1982) 'Upwelling and petroleum source beds`, AAPG Bull., Vol. 66, No. 6, pp.750-774.

Somers, M., Carson, R., Revie, J., Edge, R., Barrow, B. and
    Andrews, A. (1978) 'GLORIA 2 - An improved long range side scan
    sonar`, Oceanology International '78, Tech. Session J.,
    BPS Ltd., London.

EFFICIENT HYDROGRAPHIC SURVEYING OF EEZ WITH NEW MULTIBEAM ECHOSOUNDER TECHNOLOGY FOR SHALLOW AND DEEP WATER

R. SCHREIBER[1]
DR. H.W. SCHENKE[2]

1.   INTRODUCTION

The Multibeam Echosounder ATLAS HYDROSWEEP is designed for efficient hydrographic survey in shallow and deep water sea areas, and for the real-time display and recording of the topographical structures as coloured areas and isolines. The special features of ATLAS HYDROSWEEP are the large coverage of 2 x 45° and the shallow and deep water survey capability from 10 m to >10,000 m. With the aid of 59 PFB's (preformed beams), a swath width equal to twice the depth of the water is covered on the sea bed. This greatly increases the economy of ship operations for survey purposes, compared to previous systems. A comparison of HYDROSWEEP data with comparable SEABEAM measurements along almost identical track lines proves the high data quality and the superior efficiency of the HYDROSWEEP.

2.   OPTIMISED FREQUENCY AND TRANSDUCERS

Previous deep sea echosounders had generally been designed for 12 kHz operation. Today, the capability to reach the full ocean depth of 10,000 m or more can also be achieved with the 15.5 kHz frequency of HYDROSWEEP with its highly efficient narrow beam transducers. The low noise signal processing techniques provide a precise detection of the depth from the 59 return signals.

Another advantage of the higher frequency is the stronger "back scattering coefficient" at 15.5 kHz compared to 12 kHz. This provides better signal quality at larger angles. Also a 15.5 kHz signal penetrates less into the surface sediments, and therefore the quality of a 12 kHz return signal would be compromised by this effect. This results in an increased accuracy of the travel time measurement.

[1] Krupp Atlas Elektronic GMBH, Bremen, W. Germany
[2] Alfred-Wegener-Institute for Polar and Marine Research (AWI), Bremerhaven, W. Germany.

D. A. Ardus and M. A. Champ (eds.), Ocean Resources, Vol. I, 73–87.
© 1990 *Kluwer Academic Publishers. Printed in the Netherlands.*

Finally, the selected 15.5 kHz provides a smaller "near field", allows shorter pulses and more accurate signal threshold discrimination. All these factors contribute to the combined application for shallow water and deep sea operation.

The signal transmission is achieved by means of the two identical transducer arrays, each consisting of three modules. The arrays should be mounted in T-configuration in a protruding fairing below the base line on, or close to, the ship's centre line. An electronic stabilization assures a continuously vertical transmission of energy.

3.   ADAPTIVE TRANSMISSION BEAM FORMING

In order to reach the desired system capabilities and specifically the operational flexibility of continuous operation from 10 to >10,000 m, HYDROSWEEP offers four transmit modes in two different physical principles, omni-directional transmission (ODT) and rotational-directional transmission (RDT). The four transmit modes are automatically selected by the system software depending on the water depth.

4.   THE CROSS-FAN CALIBRATION METHOD

Gapless, accurate surveying with such a wide swath is only possible when the mean water sound velocity is known and applied for depth and slant angle corrections in real time. Therefore ATLAS HYDROSWEEP uses a unique and patented method for a quasi-real-time calibration of the measurement results. The method uses a "cross fan" configuration (Fig. 1) where the fan of 59 preformed beams (PFB's) can be generated alternatively in the direction across the vessel's axis (survey mode) or along the vessel's axis (calibration mode).

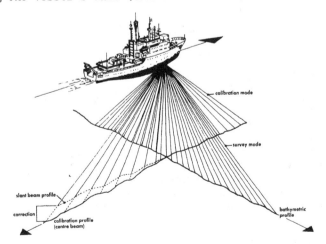

Figure 1. Principle of multibeam echosounder ATLAS HYDROSWEEP with 2 separate fans for survey mode and calibration mode.

The method is based on the consideration, that the vertical beam
(center beam) measures a true vertical travel time along the vessel's
track, while at the same profile, measured by the slant beams of the
longitudinal fan, may be deteriorated by various effects. In order to
determine the depth out of the measured travel time, the "mean water
sound velocity" (Cm) must be used in both time series. This Cm value
is then modeled during several iterations of both time series until a
value is found, which makes the "slant beam profile" fit best with the
"center beam profile" (Fig. 2).

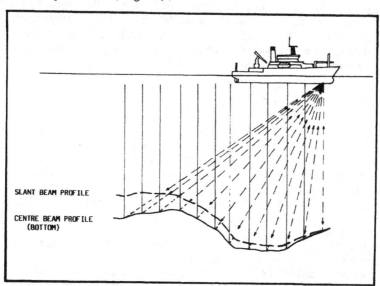

Figure 2. Measurement principle for determination of "mean water sound
velocity" (Cm)

This is determined by a "least squared residuals" comparison
between the two profiles. The calculated Cm value is then applied as
the "mean sound water velocity" to the measurements in the survey mode.

This method is very similar to the determination of RMS-Velocities
of stratigraphical layers with Normal Moveout Correction (NMO) as part
of the Common Mid Point (CMP) technique in the reflection seismics.

There are basically two effects which compromise the accuracy of a
multibeam system, specifically for systems with large swath angles such
as the 45° offered by HYDROSWEEP:

(a) The slant beams in the water column are generally bent towards the
     vertical because of refractions due to variations of water sound
     velocity. Therefore, the slant distance, measured with a slant
     beam, cannot be transformed geometrically into the vertical depth.

     A good approximation of Cm can be determined by measuring a "water
     sound velocity profile", e.g. with a CTD or XBT sonde. However,
     as such measurements are very time consuming (CTD) or costly
     (XBT), it is a great operational and also ergonomical advantage,

that for the calibration they are not necessary with ATLAS
HYDROSWEEP.

(b)   The return signal is received from the sea bed from a fairly large
      area, defined by depth, slant angle and beam width. Although the
      discrimination of the travel time may be done with utmost
      precision, the angle, from which the "energy centre" of the signal
      is received, is not known exactly. A determined correction of
      possible errors to the angle of the ray path geometry from this
      "energy distribution" effect is not possible. However, the "cross
      fan calibration method" of ATLAS HYDROSWEEP also compensates for
      the systematic part of this error contribution, because the
      longitudinal and transversal transducer arrays are identical and
      also the signal processing electronics.

5.   ECHODISCRIMINATION FOR EACH PFB

ATLAS HYDROSWEEP uses a very powerful subsystem of 8 parallel operating
microprocessors exclusively allocated to discriminate the "centre of
energy" for each PFB as the reference for the travel time measurement.
      For each channel 1-59, a memory area is filled up with the
received signal as a time series with amplitudes per time increment.
The memory now contains the shape of the bottom signal including main
echo, side lobe echoes and noise, received within the time limits of
the adaptive hardware gate.
      It is the task of the echodiscriminator to identify that part of
the time series which represents the sea bed surface.
      KRUPP ATLAS ELEKTRONIK has found a patented algorithm, which
determines the "centre of return energy" as the reference for the
travel time and depth measurement.
      Because of the various disturbing influences to the received
signal's shape a simple "peak amplitude detection" will not be
sufficient. Side lobe echoes for instance may influence the position
of the peak amplitude drastically. However, even if the peak amplitude
of a side lobe may be comparatively high, the return energy (amplitude
x time increments) is always lower (Fig. 3).

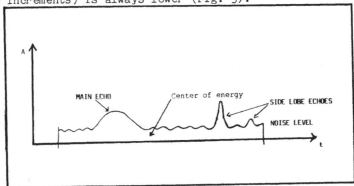

Figure 3. Determination of "Center of Energy" for Echodiscrimination
by adaptive window process.

As part of the patented algorithm, the routine to find the centre of energy is therefore repeated several times, each time with a reduced time window, symmetrically around the previously found centre of energy. By this "adaptive window process" the window is converging around the dominant centre of energy automatically.

This algorithm is repeated until the window size is corresponding to the echo length predicted for the given depth and profile shape, based on the previous measurements.

The depth discriminating method of ATLAS HYDROSWEEP is also very effective in a noisy environment. Down to a signal-to-noise ratio (S/N) of 4 dB, it is providing reliable results. The algorithm is still useful at an S/N of 3 dB where it still provides a precision of 0.45% of the depth, and it is only fading out at approximately 2 dB.

6. SURVEY RESULTS AND DATA QUALITY

This section deals with cartographic means and morphological methods of representation and scientific interpretation of multibeam data from the new HYDROSWEEP system.

Deep sea trials and especially comparisons with SEABEAM measurements were conducted on the two METEOR cruises M4/1 in 1986 and M6/4 in 1987/88. This contribution focuses on technical investigations and tests in the Gulf of Biscay and in the Romanche Fracture Zone.

The SEABEAM System on the German ice-breaking research vessel POLAR-STERN was tested and calibrated for deep sea operation in 1984 in the central and deepest part of the Romanche Fracture Zone. The same profiles were measured for comparison with the new HYDROSWEEP System, using GPS/NAVSTAR and the Integrated Navigation System INS for precise positioning.

Multibeam sonar systems have been used most successfully for scientific bathymetric surveys in deep sea areas during the last 10 years. The SEABEAM System especially, with more than 12 installations on research vessels world-wide, has been used for systematic regional as well as detailed local surveys and has provided valuable information about the seafloor topography and morphology. The technical conception of the SEABEAM System is described in detail by Renard and Allenou (1979) and the scientific use of SEABEAM in Schenke and Ulrich (1987).

A prototype of HYDROSWEEP was installed on the new German marine research vessel METEOR 1985/86. Two scientific test cruises were carried out in 1986 (METEOR leg M 4/1) and 1987/88 (METEOR leg M 6/4) for deep sea and shallow water trials.

7. NAVIGATION

Precise ship's positioning is an essential prerequisite for large scale mapping of the sea bottom topography with multibeam sonar systems such as HYDROSWEEP or SEABEAM. The survey ship has to keep its course as accurately as possible on a given track, especially for systematic box-surveys of extended areas. The parallel swath profiles should have an adequate overlap to avoid gaps in the surveyed area. The lack of land

Figure 4. Test area on METEOR cruise M 4/1, reproduced from the Carte
Bathymetrique des Canyons Sud-Americains, CNEXO 1984.

based reference points, and radio navigation systems on the oceans make
the accurate determination of a ship's positions very difficult.

Therefore the research vessels POLARSTERN and METEOR were equipped
with Integrated Navigation Systems (INDAS-V and INS) for precise
navigation. For further improvement of the position accuracy,
GPS/NAVSTAR receivers were installed onboard R/V METEOR during the
above mentioned legs.

Integrated Navigation Systems are primarily based on the Satellite
System NNSS/TRANSIT, Doppler Sonar and Gyrocompass for dead reckoning.
The accuracy from integrated systems in equatorial areas is not better
than ±300 m (Schenke and Ulrich, 1987).

Multibeam sonar survey demands precise positioning of the vessel.
Errors in the ship's position are directly transferred to positional
errors in the isolines and became apparent in a shift at the borders of
the neighbored swath tracks. Therefore it is evident that the position
accuracy must be comparable to the resolution of the multibeam sonar
system. The aperture of the preformed beams are between 1 and 3
degrees, so that the footprint diameter in 3000 m water depth would be
between 100 ... 160 m. For the planned detailed deep sea test surveys
it was indispensable to use a more precise positioning system.

The NAVSTAR Global Positioning System (GPS) is a satellite
navigation system which is currently under development by the US
Department of Defense. During this cruise, seven Block I SV's (space
vehicles) were usable. Three TEXAS INSTRUMENT TI-4100 GPS receivers
were installed onboard. TI-4100 receivers can perform P-Code distance
measurements on two frequencies L1 and L2. Two antennas were mounted
on top of the ship's main mast and one at the front of the vessel. One
receiver was linked to an on-board VAX 11/730 computer and used for
navigation. The user solution and the relative navigation information
from the TI-4100 were made available as a real-time service, both to
the bridge for accurate profile keeping and to the HYDROSWEEP Swath
Survey System. The receiver-ranging measurements data and the SV
navigation data (Ephemeris) were recorded continuously on various media
for future postprocessing. The recorded GPS data for both antenna
positions were postprocessed for optimal position references of
HYDROSWEEP soundings. The accuracy of the GPS positions is between
±20 ... 30 m.

8.   VALIDATION TESTS ON HYDROSWEEP

8.1   METEOR Cruise M 4/1

The first technical tests were performed in October 1986 on the METEOR
expedition M 4/1 in the Gulf of Biscay. Different test areas were
defined in the abyssal plain as well as at slope areas at the
continental shelf. Final postprocessing has only been carried out
until today with the data from the continental shelf. The chosen test
area (Fig. 4) was surveyed with the SEABEAM from the JEAN CHARCOT in
1978. For comparison only, the final map from these surveys could be
used (CNEXO, 1984). The HYDROSWEEP survey of this area was performed

with a large overlapping of the swath profiles. Six profiles were
measured with GPS positioning and are shown in figure 5 and six other
profiles were navigated with INS. The profile-spacing of 2 km produces
in depths of 3,500 m more than a threefold coverage of the test area.
All HYDROSWEEP measurements were gridded on the base of smoothed GPS
positions. A high precision digital terrain model (DTM) was derived
from this data set. The isoline plot of the test area is shown in
figure 5. Comparing figures 4 and 5 indicate a very good agreement
between both independent surveys.

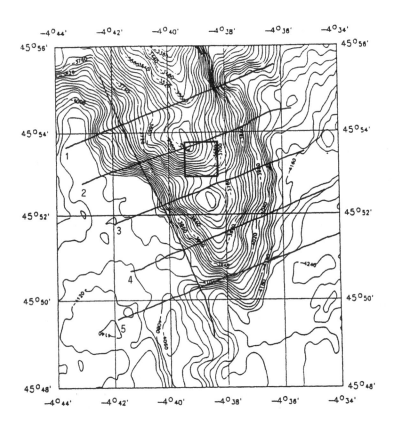

Figure 5.   Isoline map of HYDROSWEEP test area with GPS tracks.
            Scale 1 : 1,000,000, contour interval 20 m.

        Considering that the SEABEAM map is only an enlargement of the
small-scale map from CNEXO 1984, Brest, the agreement between both,
SEABEAM and HYDROSWEEP results, is surprisingly good. Caused by the
large overlapping of the profiles, the resolution of the HYDROSWEEP
survey appears better than that from the SEABEAM. However, due to the

reasons mentioned above, the topographical resolution of both maps cannot be compared directly, but the morphological structure agrees very well. Further comparison with large scale maps from the SEABEAM surveys are planned.

8.2 METEOR Cruise M 6/4

During the POLARSTERN expedition ANT III/1 in 1984 extended calibration measurements were performed in the deepest and steep part of the Romanche Fracture Zone (Schenke 1985). The Romanche Fracture Zone is the largest and most rugged trench of the Mid Atlantic Ridge with extreme steep flanks and slope inclinations of more than 40°. This area was also surveyed from the JEAN CHARCOT in 1977 (Allenou and Renard, 1979). The SEABEAM survey of POLARSTERN was performed with the Integrated Navigation System INDAS-V only. The positional accuracy can therefore only be expected between ±300 ...400 m. This survey was performed with a small overlaps of about 5%. The method of the postprocessing of navigation and SEABEAM data is described in (Schenke and Ulrich, 1987). The results of the SEABEAM survey is shown in figure 9 in the form of an isoline plot and in figure 11 as a three-dimensional plot.

After the first successful scientific tests during leg M 4/1, the German Science Foundation and the Oceanographic Science Board decided to perform additional tests with the new HYDROSWEEP onboard METEOR and to compare these measurements to the SEABEAM data of POLARSTERN.

The HYDROSWEEP survey at Romanche Fracture Zone was carried out in order to prove the performance of the new HYDROSWEEP under difficult conditions in rugged submarine topology on the METEOR cruise M 6/4 from December 28, 1987 until January 12, 1988. Exactly the same profiles from the SEABEAM survey 1984 were sailed twice in opposite directions, one with integrated navigation only and the second time with high precision GPS navigation to test the capability of the Integrated Navigation System INS. GPS and INS survey was performed alternatively with respect to the GPS observation windows. The whole survey of the R.F.Z. could be maintained with the full swatch capability of HYDROSWEEP (90°) with a ship's speed of about 10 kts.

After the postprocessing and combination of navigation and multibeam data and diligent error snooping, the internal data quality was tested and analyzed in a first step under the following different conditions:

- Data quality on extremely steep slopes.

- Test of the maximum water depth at 8,000 m and the entirety of the fan.

- Analysis of adjacent tracks in opposite directions, at uphill and downhill overlaps.

- Comparison of center beams (higher accuracy) with outermost beams (lower accuracy).

The statistical analysis of the HYDROSWEEP data was investigated by means of sophisticated statistical programs. Figures 6 and 7 show typical graphs for the distribution of standard deviations (S.D.) versus swath width. The standard deviations were determined on the basis of one profile across the R.F.Z. (Fig. 6) and of all profiles (Fig. 7). They only represent the internal precision of the measured data. The curves confirm the accuracy for the HYDROSWEEP System, specified by the manufacturer. They also agree with results of an accuracy study in Tyce and Ziese (1987).

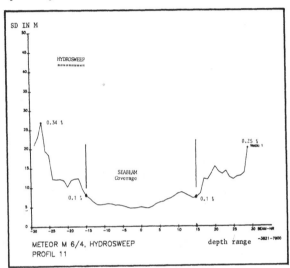

Figure 6.   Distribution of standard deviation versus swath width. Representative data from HYDROSWEEP profile No. 11.

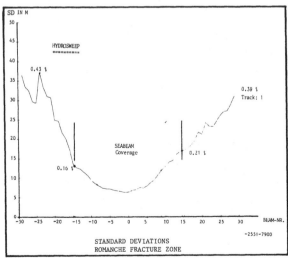

Figure 7.   Distribution of standard deviation versus swath width. Representative data from all HYDROSWEEP profiles.

A typical swath plot presenting two overlapping HYDROSWEEP stripes at the southern slope, covering a depth range between 4,000 and 7,700 m of the Romanche Fracture Zone, is shown in figure 8. This result demonstrates impressively the entirety of the data and the conformity of adjacent swaths.

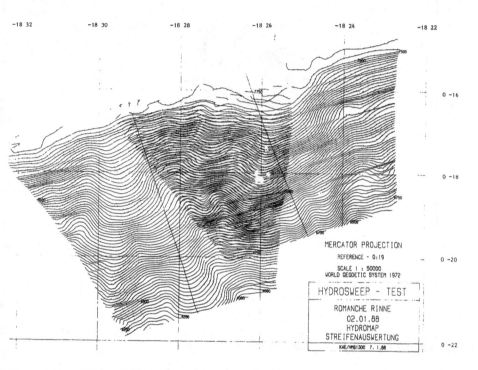

Figure 8. HYDROSWEEP swath plots from the south slope of Romanche Fracture Zone with high overlapping.

In the second stage of the postprocessing, a digital terrain model (DTM) was determined. For the avoidance of artifacts in the final DTM spurious and erroneous measurements had to be identified and deleted. 3% of the entire data had to be edited manually. The whole data set of the 1,200 km^2 large test area in the R.F.Z. contains approximately 220,000 data points which gives a mean point distance of about 70 m. The number of data points of the 900 km^2 large SEABEAM test area was only 120,000 which gives a mean point distance of 85 m. The isoline representation of the R.F.Z., computed with the program system TASH (Kruse, 1987) from SEABEAM data (Fig. 9) and from HYDROSWEEP data (Fig. 10) indicates a very good agreement between both maps.

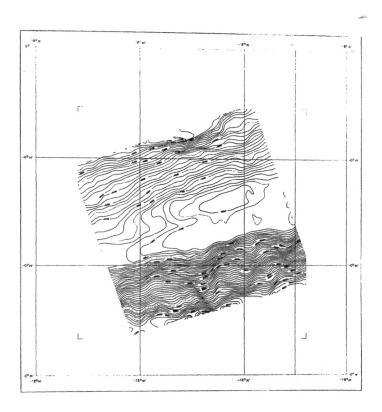

Figure 9. Final Bathymetric Chart of the Romanche Fracture Zone, based
on the SEABEAM survey, contour interval 100 m.

        Because of the high overlapping of the profiles, a grid distance
of 100 m could be used for the DTM-determination. Digital Terrain
Models may be used beside isoline interpolation also for computation of
perspective block diagrams. Figures 11 (SEABEAM) and 12 (HYDROSWEEP)
demonstrate the derived products. The depths of the approximately
30 x 40 km² large test area are between 3,000 m on the south shoulder
and 7,850 m in the trench of the R.F.Z., and slope gradients up to 45%
inclination can be seen. Figures 9 and 10 show that the morphological
structures are well represented by both survey results. Significant
discrepancies are not detectable. The standard deviations derived from
the DTM-determinations are about ±50 m for both systems.

9.   CONCLUSIONS

HYDROSWEEP, the new German 90 degrees fan-shaped multibeam system with
59 preformed beams, installed on the new R/V METEOR, has been proved as
a powerful bathymetric survey system for deep and shallow water
operation. In the frame of two scientific test cruises with the
METEOR, direct comparisons with SEABEAM products were performed. From

these comparisons it was possible to show that the results of
HYDROSWEEP surveys are at least of the same quality and reliability.
During the test cruises it was shown that HYDROSWEEP could operate with
the entire fan in water depths of 8,000 m. Generally, the internal
accuracy is 0.5% of the water depth along a HYDROSWEEP track. This
value is too optimistic. The standard deviation derived from the DTM-
determination is approximately 1% of the water depth. However, the
manufacturer's specification for the internal accuracy of 1% over the
entire fan is also valid for a digital terrain model derived from a
large area box survey, which contains also the positioning errors as a
strong limiting factor. Therefore high precision bathymetric
HYDROSWEEP surveys should be performed with GPS/NAVSTAR positioning
only.

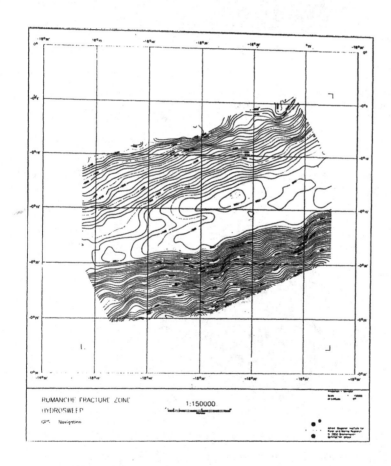

Figure 10. Final Bathymetric Chart of the Romanche Fracture Zone,
based on the HYDROSWEEP srvey, contour interval 100 m.

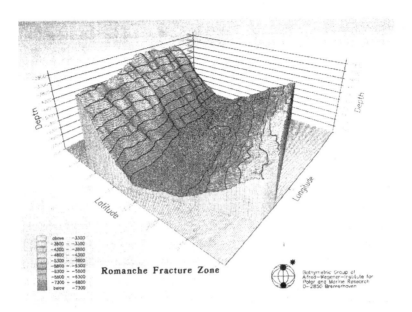

Romanche Fracture Zone

Figure 11.   Perspective view from the North-East into R.F.Z., based on
a DTM from SEABEAM data.

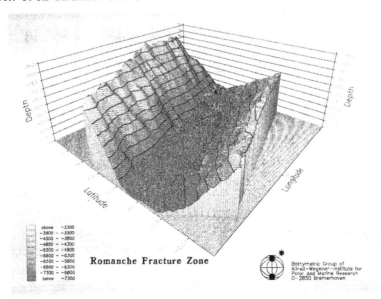

Romanche Fracture Zone

Figure 12.   Perspective view from North-East into R.F.Z. based on a DTM
from HYDROSWEEP data.

10. ACKNOWLEDGEMENTS

The scientific test cruises METEOR M 4/1 and M 6/4 were supported and financially sponsored by the German Science Foundation under Grant DFG He 89/52-1 and DFG Sche 296/1-1/2. We than Professor Seeber, Institute of Geodesy, University of Hannover, for the fruitful co-operation. We extend our thanks to the following institutions for providing GPS receivers: Professor H. Pelzer, Geodetic Institute of the University of Hannover; Professor R. Rummel, University of Delft, The Netherlands.

11. REFERENCES

'ATLAS HYDROSWEEP Hydrographic Multibeam Sweeping Survey Echosounder, Operating Instructions' (1987) Krupp Atlas Elektronik, Bremen.

'ATLAS HYDROSWEEP Hydrographic Multibeam Sweeping Survey Echosounder, System Description 9/88' (1987) Krupp Atlas Elektronik, Bremen.

'Carte Bathymetrique des Canyons Sud-Americains' (1984) Centre National pour L'Exploitation des Oceans (CNEXO).

Kruse, I. (1987) 'TASH - Ein Programm-System zur Berechnung von Digitalen Gel¨ndemodellen (DGM) und zur Ableitung von Isolinien' in 4. Kontaktstudium des Instituts für Kartographie der Universität Hannover.

Renard, V. and Allenou, J.P. (1979) 'Multibeam Echosounding on JEAN CHARCOT, Description, Evaluations and Results' in The International Hydrographic Review, Vol. LVI(1), Monaco, pp.36-67.

Schenke, H.W. (1985) 'Kalibrierung und Erprobung der SEABEAM-Anlage bei großen Wassertiefen' in Die Expedition ANTARKTIS III mit FS "POLARSTERN" 1984/85, Herausgeber: Gotthilf Hempel, Berichte zur Polarforschung Nr. 25, ISSN 0167-5027.

Schenke, H.W. and Ulrich, J. (1987) 'Mapping the Seafloor', Applied Geography and Development, Vol. 30, ISSN 0173-7619.

Schenke, H.W. (1987) 'Ergebnisse und Analysen bathymetrischer Vermessungen mit Fächersonarsystemen', Beiträge zum 3, Hydrographie-Symposium der DHyG, Bremerhaven.

Tyce, R.C. and Ziese, R. (1987) 'Genauigkeitsanalyse von Fächersonar-Ergebnissen', in DHyG-Information, Heft Nr. 009, Herausgeber, Deutsche Hydrographische Gesellschaft, Wetternstr. 8, 2160 Stade, ISSN 0934-7747.

ENGINEERING SOLUTIONS FOR DEEPWATER FOUNDATION PROBLEMS USING
INTEGRATED INVESTIGATIONS

JAMES R. HOOPER AND ALAN G. YOUNG
Fugro-McClelland
Houston
Texas
USA

ABSTRACT. Frontier regions of mineral exploration present engineers
with foundation design problems that are outside the bounds of normal
experience. To understand the risk of economical development, geologic
and engineering models of the site and region must be developed.
Through integration of the models, the interaction between geologic
processes and engineering parameters can be clarified and the effects
of construction can be assessed. An example study of a site
investigation in 1,800 ft of water on the continental slope in the
northern Gulf of Mexico is presented. Geologic studies included
sedimentology, geochemistry and interpretation of geophysical seismic
data. Cone penetration and temperature measurements were performed in
a boring which penetrated to 652 ft below the seafloor. The results of
analyses of geologic and engineering models suggest that small
quantities of gas hydrates exist as thin layers disseminated in the
ground about 300 ft below the site and that the process of hydration is
continuing. This knowledge resulted in design alterations to mitigate
the effects of gas hydrates on foundation performance.

1.   INTRODUCTION

The continental slope in the Gulf of Mexico (Fig. 1) is a frontier
province for foundation engineering. Soil strata have developed by
geologic processes that are unfamiliar to foundation engineers in
comparison to conditions on the adjacent shelf. There is a
possibility, therefore, that novel foundation design problems may be
encountered.
        Recent studies on the slope have combined engineering measurements
of soil properties with comprehensive geologic analysis. Relationships
between formative processes and the engineering properties of the
foundation materials were defined. This integrated approach to
foundation investigations minimizes unpleasant surprises during
installation and operation of a deepwater facility. In the following
discussion an example is presented to show how engineering and geologic
techniques of analysis were combined to illuminate an unusual
foundation problem on the continental slope in the Gulf of Mexico.

89

D. A. Ardus and M. A. Champ (eds.), Ocean Resources, Vol. I, 89–101.
© 1990 Kluwer Academic Publishers. Printed in the Netherlands.

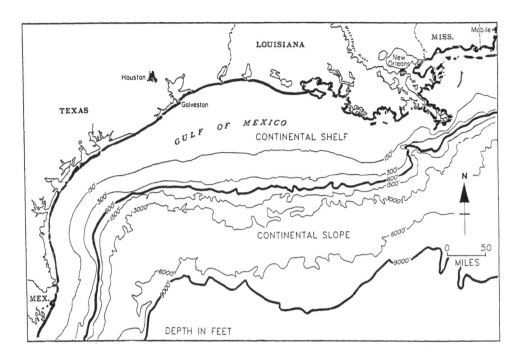

Figure 1. Continental Slope Northern Gulf of Mexico

2. GAS HYDRATES IN FOUNDATION STRATA

A tension-leg platform production facility was planned for a site on
the continental slope offshore Louisiana. During preliminary studies,
drop core samples were obtained from the top of a large mound on the
seafloor about a mile from the site. The samples contained fragments
of an ice-like solid which quickly melted and expelled gas vapor. The
fragments were identified as methane hydrate. Since oil and gas
production would inevitably heat the surrounding soil, there was
concern that if hydrates were present beneath the site, they would
melt, release gas into foundation strata and reduce the stability of
the structure.

2.1 Characteristics of Gas Hydrates

When certain types of solid gas molecules are in the presence of water
at low temperature and high pressure, a solid crystalline substance
called a "gas hydrate" may result. The basic crystal unit is a
spherical-shaped lattice structure or "clathrate" of water molecules
with interior spaces that accommodate gas molecules. A hydrate
composed of clathrate units which are completely filled with methane
gas molecules may contain as much as 170 unit volumes of gas per unit
volume of hydrate (Kuustraa et al, 1983). This gas is released when
the hydrate melts.

The physical appearance of methane hydrate is similar to that of normal ice. When it occurs within sediments, it may take on physical forms ranging from layers tens of feet thick to small crystals which are disseminated within a soil matrix (Brooks and Bryant, 1985). Physical properties of hydrocarbon gas hydrates such as thermal conductivity, heat of crystallization, etc., resemble those of normal-water ice but depend upon the chemical composition of the gas and liquid (Kuustraa et al, 1983) (Miller, 1974).

Pressure-temperature relationships for methane hydrate and for methane mixed with other hydrocarbon gases are shown on figure 2. Hydrates may form at temperatures and pressures to the left of a phase line; only gas vapor exists in the presence of water to the right of a line. Small amounts of propane or ethane with methane will initiate freezing at temperatures that are warmer than those required for pure methane at the same pressure. Conversely, salt in pore water has the effect of lowering the freezing temperature of hydrates (Miller, 1974) (Macleod, 1982). Both natural biogenic and reservoir-formed hydrocarbon gases often exist as mixtures of methane and other hydrocarbon gases, but reservoir derived gases may contain large percentages of the heavier hydrocarbons.

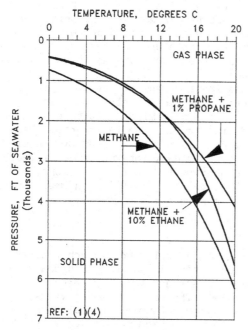

Figure 2. Phase diagrams for gas hydrates.

2.2 Gas Hydrates in the Gulf of Mexico

The existence of gas hydrates in the Gulf of Mexico was confirmed by Brooks and Bryant in 1985. They studied eight locations offshore Louisiana and Texas using high resolution geophysics and soil sampling techniques. Hydrates were recovered from seafloor deposits and from strata about 100 ft below the seafloor. Four of the locations contained methane hydrates derived from biogenic methane. The other four were from thermogenic (reservoir) hydrocarbon gases which had seeped upward from depth. The water depths of the biogenic-derived hydrates ranged from 2,800 to 7,800 ft. Thermogenic gas hydrates were encountered in water depths of 1,700 to 4,200 ft.

Seafloor temperatures on the slope offshore Louisiana and Texas may range from 7 to 15°C in 1,500 ft water depth (Dept. of Navy, 1967). In areas characterized by the colder temperatures, seafloor accumulations of hydrates theoretically can occur in water depths as shallow as about 1,500 ft. The lower boundary of the zone of hydrate formation will be determined by geothermal warming of sediments at depth. In 1,500 ft of water the frozen zone would be limited to a layer near the seafloor. Further down the slope where the pressure is greater and the water is colder, gas hydrates may form in sediments buried deep below the seafloor. The zone of potential hydrate formation is more than 1,000 ft thick in deeper waters.

2.3 Preliminary Assessment of Foundation Gas Hydrates

The proposed structure site is on a gently sloping featureless seafloor in a water depth of about 1,800 ft. A geophysical seismic profile across the region is shown on figure 3. Based on seismic data, geophysicists defined 5 stratigraphic units. The study did not show any particularly distinctive reflection patterns which would indicate the presence of gas hydrates.

Previous sample borings in this general region of the slope had shown that the stratigraphic units were likely to be massive clay sections of low permeability with only minor amounts of granular materials such as silt and fine sand. This is important because the growth of hydrate layers might possibly be limited by the slow transmission of gas vapor. Preliminary information from geophysics and regional knowledge of sediment type indicated, therefore, that if hydrates were present at the site, they might be in small amounts scattered throughout the sediment profile.

2.4 Field Exploration

The exploration program combined gelogic and geotechnical engineering techniques to detect possibly minor amounts of gas hydrates within the sediment column. Four borings were drilled at corners of a square pattern about 200 ft on a side. The primary boring was sampled to a depth of 652 ft below the seafloor. Samples were also recovered from a second boring to a depth of 272 ft. Each sample was about 2-ft long and was obtained at 10-ft intervals to 500 ft and at 20-ft intervals below that to the bottom of the boring. Samples were examined and tested to determine geologic characteristics and engineering properties. Small portions of each sample were sealed in containers for geochemical analysis of gas and pore fluid.

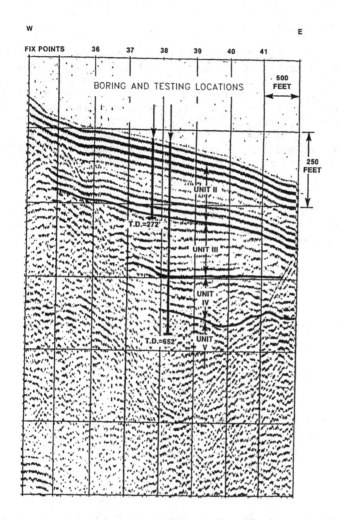

Figure 3   Geophysical profile

The two remaining borings were used for measurements of engineering properties from the seafloor to a depth of 300 ft. These measurements included in situ strength testing using a cone penetrometer. Cone penetrometer tests are hydraulically controlled thrusts of a rod for distances of 10 ft below the bottom of the borehole. A cone test was performed every 20 ft. Instrumentation in the tool measured the resistance to penetration of the tip and of a sleeve on the side of the cone. Frozen zones (hydrates) are expected to be much stronger than unfrozen.

The temperature of the sediment was obtained by pushing a thermister probe below the bottom of the primary (deepest) borehole. The purpose of these measurements was to determine the temperature regime, allowing analysis of gas hydrate phase conditions.

2.5 Stratigraphic Profile

Significant stratigraphic characteristics of samples are presented on figure 4. No gas hydrate ice crystals were found in any of the samples after recovery to the deck of the drill ship. However, thin zones of a weak, watery and gassy clay were noted at depths ranging from 270 to 370 ft. The layers 1- to 2-inches thick, were described as having the consistency of soft ice cream. The clay on top and bottom of the layers was intact and strong. It is possible, therefore, that the weak layers resulted from melting of gas hydrates during the 15-20 minutes required to bring the sample tube from the bottom of the borehole to the deck.

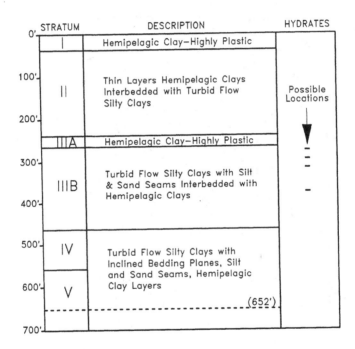

Figure 4. Description of sediments

     Many samples between 50 and 550 ft showed traces of oil both by smell and by visual observations. Oil was not consistent in its presence but varied in quantity both vertically and horizontally between borings. A small amount of gas vapor was encountered near 270 ft in one of the borings.
     Specimens from the borings were all clays of high to moderately high plasticity. When samples were studied using x-ray radiography, it was found that stratigraphic (geophysical) units I and IIIA originated from hemipelagic deposition, a process of slow accumulation of particles from out of the water column. Units II, IIIB, IV, and V contained thin hemipelagic layers interbedded within siltier clays which had derived from mass wasting processes such as turbidity flows.

Thin (mm-thick) lenses and pockets of silt occurred in samples within some zones and a few carbonate pebbles were found near 270 ft. In addition, Unit V was typified by inclined bedding planes which may be the result of slumping, debris flows or similar movements during the geologic past.

Hemipelagic clays in this region of the Gulf of Mexico have low permeability to water and gas. However, silt and sand lenses may accumulate gas vapor. Therefore, gas hydrates would be more likely in deposits such as Units II, IIIB, IV, and V which were derived from mass wasting processes conducive to development of granular layers with locally high permeability.

2.6 Geochemical Characteristics

Light hydrocarbon gas analysis was performed on the pore vapor and pore water of clay specimens that were sealed in containers immediately after recovery of the sample to the deck of the drilling vessel. The samplers were not pressure sealed and some gas was undoubtedly lost from the samples while coming up the drillstring. However, the relative amounts of different gases were probably represented in the tests.

Geochemical analyses found significant concentrations of hydrocarbon gases from 40 ft to about 550 ft. The gas was identified as methane mixed with small amounts of other hydrocarbon gases. These allied gases (propane, ethane, etc.) were about 1-2% of the total volume of gas within the depth range from 300 to 350 ft. Generally, hydrocarbon gases other than methane were only about 0.2 to 0.5% of the total for depth ranges from 40 to 300 ft and 350 to 550 ft.

Testing indicated that the largest concentration of oil occurs between 270 and 310 ft but traces were found in many other samples. Further geochemical testing showed that the oil in the samples was biodegraded and that the process of biodegradation was the primary source of the hydrocarbon gases. The analysis also suggested that a minor portion of the gas was from reservoir leakage, but the evidence was inconclusive and it seemed more likely the gas was biogenic in origin.

2.7 Cone Penetrometer Resistance of Sediments

The cone penetrometer was pushed through a 10 ft section of sediment out of every 20 ft down the borehole to a depth of 300 ft. In the two in situ borings, cone data was collected in overlapping zones so that complete coverage was obtained. From the seafloor to about 230 ft the resistance to penetration increased linearly, reflecting a normal pattern of strength increase with depth in clays. From there to the bottom of the tested zone (300 ft), however, the resistance exhibited considerable variation, increasing and decreasing about 10% from the norm. An exception was at 265 ft where the cone encountered a high resistance object that was only penetrated with great difficulty. The hard zone was about 1-ft thick.

One possibility for the variations of cone resistance is that the tip penetrated thin layers of strong sediments within the turbid flow strata of Units II and IIIB. This result is possible for strata derived from mass movements. The large resistance at 265 ft might be caused by a thin zone of small pebbles. Carbonate pebbles were found in one of the sample borings near this depth. There is also the possibility, however, that the variations in penetration within the soil profile indicates gas hydrates.

2.8 Temperature Regime

Data from downhole temperature measurements are shown on figure 5. Phase relationships for gas hydrates of pure methane and 99% methane-1% propane are also shown on the profile for comparison. The seafloor temperature of 7°C is low enough to encourage freezing of both mixed and pure hydrocarbon gases. Temperatures below 160 ft are all warmer than the phase line for pure methane hydrates but are cold enough for the formation of hydrates of mixed gases. It should be recalled that geochemical tests showed methane is mixed with 1-2% other hydrocarbon gases in samples recovered from 300 to 350 ft.

The thermal gradient line A on figure 6 was drawn from the measured temperature at 652 ft to the seafloor measured value. A linear geothermal gradient of this type would result if heat was flowing through an idealized sediment of uniform thermal conductivity. Line B on figure 6 is nonlinear. It was calculated by heat flow theory

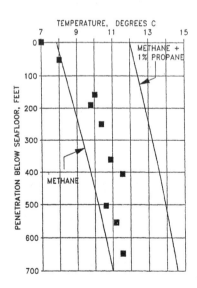

Figure 5. Temperature measurements and phase lines.

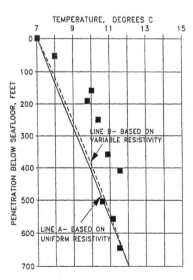

Figure 6. Temperature measurements and geothermal gradients.

using thermal conductivity values at the site derived both from test
measurements and from correlations with sediment properties such as
density and water content. Differences between the two gradient lines
are small, less than 0.2°C at any depth. Temperature measurements
between 160 ft and 410 ft, on the other hand, are as much as 1.5°
warmer than the theoretical geothermal gradients. It is difficult to
explain this as an effect of layered variations in thermal
conductivity.

One possibility for the apparent temperature anomalies is that the
temperature is controlled by present-day hydration within the sediment.
In the model on figure 7, it has been assumed that gas hydrates are
forming deep below the ground at small nucleation sites distributed
within a layer of sediment. During freezing, the temperature at each
spot in the layer will correspond to the phase line for the particular
mixture of gases. The rate of freezing will be controlled in part by
the availability of hydrocarbon gas at the freezing front.

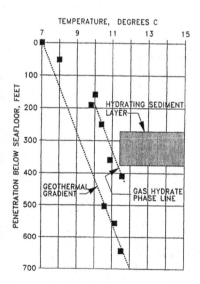

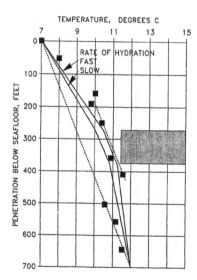

Figure 7. Stratigraphic model for   Figure 8. Temperature profile for
hydration analysis.   hydrating layer.

A second major control on hydration rate is the comparatively low
thermal conductivity of the sediments, because the heat of
crystallization must be supplied to the freezing front by conduction
through the ground. One effect of rate of freezing of gas hydrates on
the temperature profile is suggested on figure 8. If a high rate of
freezing can be maintained, then the average temperature within a layer

of sediment will be that corresponding to the phase line condition.   If
the rate is slower, then the average temperature will be less.
     A specific physical model was numerically analyzed to determine
the potential effect of on-going hydration on the temperature profile.
It was assumed that freezing was presently occurring within a layer of
clay sediments about 100-ft thick, centered about 320 ft below the
seafloor. The gas is assumed to be produced continously from the oil
that has penetrated the region. Nucleation sites are located in thin
layers and lenses of silt that are scattered throughout the turbid flow
layers. In the study, temperature vs. depth was calculated for several
different rates of hydration, producing profiles similar in shape to
figure 8.
     The average temperature vs. time in the center of the freezing
zone is shown on figure 9, expressed in terms of the rate of heat
transmission to the layer. If the heat of crystallization rate is 8
calories per cubic ft of sediment per year (cal/cuft yr), then the
sediment layer temperature could be maintained about 1.5°C above that
of the normal geothermal gradient. The analysis also indicates that
steady state equilibrium would be established about 1,000 years after
the process begins within the layer.

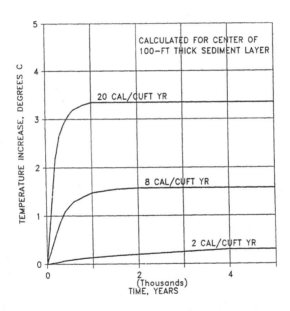

Figure 9. Temperature increase in hydrating layer.

     Conditions at the site of the boring are likely to differ
significantly from the simple model used in the numerical analysis. If
hydration is actually occurring, the sites of nucleation within the
ground probably vary laterally and vertically, as opposed to uniform 2-
dimensional assumptions of heat flow theory. Also, gas mixtures
probably vary in quality throughout the sediment column such that
different phase lines may apply from place to place. This may explain

the rather irregular temperature profile measured in the borehole. The analysis does tend to support the hypothesis that gas hydrates are actively crystallizing within a layer deep below the seafloor.

It is interesting to note that at a hydration rate of 8 cal/cuft yr, something in the vicinity of 0.01 cubic inches of hydrate per cubic ft of sediment would be produced per year. If we presume that the hydration process started near the end of the last major low stand of sea level during the Late Pleistocene, 15,000 to 25,000 years ago, and has continued ever since, the present day volume of ice would be about 10% of the total volume of sediment in the active layer. Based on other evidence (cone penetration profiles and sampling), this is also the maximum amount of gas hydrate that would be predicted at the location.

2.9 Summary

The convergence of circumstantial information from geologic and engineering sources leads to the conclusion that gas hydrates probably exist in the strata beneath this site. If so, they only occur in thin layers which are accumulating very slowly in small pockets and lenses of silt where gas vapor can form. The zone of hydration may be limited to a layer of 100-ft thickness located several hundred feet below the seafloor. Some of this limited volume of gas hydrate will melt when heated by oil and gas production in wells. Based on this model, the foundation design was adjusted to mitigate any significant effect of soil temperature change on the stability of the structure including melting gas hydrates.

3. COMBINING RESOURCES

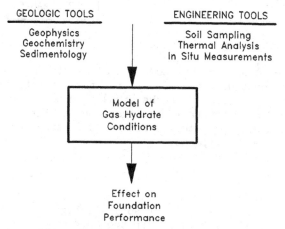

Figure 10. Integration of engineering and geology for gas hydrate investigation.

This example of a foundation investigation in a frontier environment is summarized on figure 10. Geologic and engineering tools of analysis were combined to develop a model of gas hydrate conditions in the ground beneath a proposed structure on the continental slope. Using the model, engineering studies were performed to determine the effect of gas hydrates on foundations and to design safe foundation elements.

The study of gas hydrates was only one aspect of the investigation for this particular site. Considerations were also given to a wide range of potential problems, regional and site specific, related to slope stability, active faulting, large carbonate hills, sensitive deposits of clay at the seafloor and others that are peculiar to the continental slope in the Gulf of Mexico. As suggested on figure 11, the combined efforts of foundation engineers and geologists are integrated to form models which can be used to study the impact of geological conditions on the full range of regional and site-specific problems of engineering design.

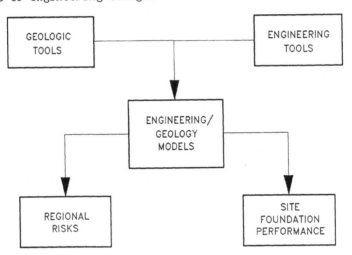

Figure 11. Project integration concept.

The deepwater slopes of the continental margin are frontier environments for construction of facilities that will be required to develop offshore resources. When the study team combines the talents of professionals from a spectrum of views, then an integrated model can be developed which will provide the fullest understanding of the problems of resource development.

4. REFERENCES

Brooks, J.M. and Bryant, W.R. (1985) 'Geological and geochemical implications of gas hydrates in the Gulf of Mexico', Rpt. to DOE Morgantown Energy Tech. Center, Morgantown, WV, p.131.

Dept. of Navy (1967) 'Oceanographic atlas of the North Atlantic Ocean, Section II, Physical properties', US Naval Oceanographic Office, Washington DC, p.300.

Kuustraaz, V.A., Hammershaimb, E.C., Holder, F.D. and Sloan, E.D. (1983) 'Handbook of gas hydrate properties and occurrence', Rpt. DE84-003080, DOE Office Fossil Energy Tech. Center, Morgantown, WV, p.234.

Miller, S.L. (1974) 'The nature and occurrence of clathrate hydrates', in Natural gases in marine sediments, Ed. I.R. Kaplan, Plenum Pub. NY, pp.151-177.

Macleod, M.K. (1982) 'Gas hydrates in ocean bottom sediments', Amer. Assoc. Petrol Geol. Bull., 66(12), pp.2649-2662.

PART III

Geological Utilization

OVERVIEW OF MINERAL RESOURCES IN THE EEZ

D.S. CRONAN
Marine Mineral Resources Programme
Imperial College
London
U.K.

ABSTRACT. There are at least six classes of EEZ mineral deposits, aggregates of sand and gravel, metallic placer minerals, phosphorites, black smoker related polymetallic sulphides and associated metalliferous sediments, manganese nodules and cobalt-rich manganese crusts. Of these, aggregates and placers are shallow water deposits and have been mined for several decades and whose extraction poses few technological difficulties other than those associated with mining in progressively deeper or more hostile environments. Phosphorites are mid-depth deposits, generally not more than a few hundred metres deep, which have yet to be mined offshore. Prospective deposits such as those on Chatham Rise off New Zealand are coarse gravels 0.8-5.0 cm in size. Proposed mining techniques involve a hydraulic lift system incorporating various separators, as yet still at the drawing board stage. Black smoker deposits are composed principally of Cu, Zn and Fe bearing sulphide minerals which precipitate from hydrothermal solutions discharging at high temperature onto the ocean floor. Most occurrences are at mid-ocean ridge crests at around 3500 m depth, but they also occur in island arcs, sometimes at shallower depth. The nature of the deposits depends on many factors including their temperature of formation, substrate rock composition and rock-water reactions taking place below the sea floor. Resource feasibility studies on these deposits have only been carried out in the Red Sea where in the 60 km² Atlantis II Deep there are 30 million tonnes (m.t.) of Fe, 2 m.t. of Zn and 0.5 m.t. of Cu, together with significant amounts of silver. A pre-pilot test mining operation has been carried out there. Manganese nodules are mainly non-EEZ deposits being concentrated in the proposed future International Seabed Area (ISA) of UNCLOS, more than 200 miles from land. However, important nodule deposits occur within the EEZ's of some Pacific island nations such as the Cook Islands and on the Blake Plateau in the United States EEZ, which might be mined preferentially to the ISA ones. The largest deposits are rich in Co rather than in Ni and Cu, the two most economically important metals in the ISA. Cobalt-rich manganese crusts have been known for almost as long as the nodules, but have only recently been thought of as an economic deposit. They are attached to hard rock substrates mainly on seamounts between 800-2400 m depth and pose considerable mining

D. A. Ardus and M. A. Champ (eds.), Ocean Resources, Vol. I, 105–111.

problems in that, unlike nodules, they would have to be broken off
before recovery. Average Co in the deposits is around 1%, but
recovered ore will contain less Co than this due to dilution by
substrate material and may be no richer in Co than the 0.5% Co typical
of the Co-rich nodules mentioned above. Recent resource assessments
for Co in both nodules and crusts in the EEZ's of the Cook Islands and
parts of Kiribati have shown much greater quantities of Co in the
nodules than in the crusts, suggesting that the relative resource
potentials of these deposits in EEZ's needs to be reassessed.

1.   AGGREGATES AND PLACERS

Aggregates are quantitatively the most important marine mineral deposit
being mined at the present time, and marine aggregates account for as
much as 15% of the total aggregate production in the U.K. Japan is
also an important offshore aggregate producer, mining over 40 m.t. a
year. Most of the mining is carried out by hopper or trailing
dredgers, which operate on a quick turn around in port. Work in the
European EEZ is mainly done down to 35 m depth, but depths of 40-50 m
are worked off in Japan.
      Placers are detrital metallic minerals found preferentially on
beaches and in near shore areas. Historically important deposits
include the tin deposits of S.E. Asia, the placer mineral being
cassiterite, and the diamonds off S.W. Africa. However, quantitatively
more important are placer deposits of light heavy minerals such as
zircon, ilmenite and rutile, which occur mainly on mid-latitude high
energy beaches.
      Placer mineral deposition represents a balance between the supply
of the minerals from the land and the energy levels in the nearshore
environment needed to concentrate them into workable deposits. With
variations in sea level over the past few hundred thousand years,
placers are found on both raised and submerged beaches. When a placer
mineral bearing beach is submerged by rising sea level, one of two
things can happen to the deposit. Either the advancing sea can
transport the minerals shorewards to concentrate them on the modern
beach, as has happened off eastern Australia for example, or it can
disseminate the minerals over a wider area than that of the pre-
existing beach, as has happened off Mozambique.
      The recent increase in the price of noble metals has stimulated
much noble metal placer exploration. This is most evident in the case
of gold, but is reflected in increased platinum exploration too.
Previously uneconomic gold deposits off Nome, Alaska, are now being
mined, and there is a considerable interest in gold off the Solomon
Islands and in other Pacific island EEZ's.

2.   PHOSPHORITES

Phosphorites are calcium phosphate minerals whose main use is as an
agricultural fertilizer, although the mineral also has uses in the
chemical and pharmaceutical industries. They occur in two main

settings, off western (and sometimes eastern) continental margins, and on oceanic seamounts.

Many continental margin phosphorites are formed as a result of oceanic upwelling supporting high biological productivity in the surface waters, thus ensuring a high flux of sinking organic remains to the seafloor. There they undergo decay, liberating phosphorous into the interstitial waters of the sediments to build up in concentration and ultimately to precipitate as phosphate minerals.

Seamount phosphorites are of more uncertain origin. Some have undoubtedly formed from bird droppings when the seamounts were islands in the geological past, but many are older than the geological age when birds originated and thus cannot have formed in this way. From the economic point of view, seamount phosphorites are less important than continental margin varieties, but they could be locally important for island nations who currently import phosphorite from abroad.

One phosphorite deposit that has generated considerable interest in recent years is that on the Chatham Rise off New Zealand. Agricultural experiments have shown that it can be applied to soils directly, without expensive pre-treatment. The deposit consists of a loose nodular gravel up to 70 cm thick, with individual phosphorite lumps ranging in size from a few millimetres to about 15 cm across. The phosphate content of the deposits is variable but averages about 9.4% elemental P, which is slightly higher than the P content of commercially available superphosphate fertilizer. Fluorine and uranium in the deposits, potentially important by-products, are also quite high, and neither detract from the fertilizer properties of the phosphorite. Other continental margin phosphorites of potential future economic importance are off the Carolinas and Baja California. Although close to onshore deposits, these may ultimately be mined as a result of land use restrictions onshore due to environmental or land use constraints.

Phosphorite mining will involve several steps, including uptake of the phosphorite bearing sediments from the seafloor, separation of phosphorite lumps, lifting the ore to the ship, and disposal of accompanying barren sediments. All this will cause considerable disturbance of the benthic ecosystem, including resuspension of sediment and the loss of growth sites for organisms on the phosphorite nodules. Much work needs to be done on both mining systems and on minimising environmental disturbance before offshore phosphorite mining can become a reality.

3.   POLYMETALLIC SULPHIDES AND METALLIFEROUS SEDIMENTS

Polymetallic sulphides and their associated metalliferous sediments consist of sulphides of Fe, Cu and Zn, oxides and silicates of Fe and oxides of Mn, formed by hydrothermal processes resulting from seafloor volcanic activity. They have been mainly found on mid-ocean ridge spreading centres, including the Red Sea, but also occur in island arcs. Most of the mid-ocean ridge deposits occur in the International Seabed Area, but the island arc and Red Sea deposits fall in EEZ's.

The deposits are formed as a result of seawater entering sub-seafloor rocks through cracks and fissures, leaching metals from the rocks, and becoming transformed into a mineralising solution. This then rises back to the seafloor to discharge as a "black smoker" which is a jet of finely divided sulphide precipitates. Black smokers discharge through "chimneys" which can reach heights of several metres before they collapse to form mounds of broken up sulphide chimney material. The accumulation and coalescence of "chimney mounds" can lead to large deposits of sulphide minerals analogous to massive sulphide ore bodies on land. Some mineral precipitation also takes place below the seafloor, giving rise to a sulphide "stockwork" in the upper part of the oceanic crust, similar to that under many on-land sulphide deposits.

The composition of polymetallic sulphides and metalliferous sediments is highly variable and depends on several factors including temperature, the composition of the sub-seafloor rocks being leached, and the rock-water ratio at the site of leaching. Of these, the composition of the sub-seafloor rocks can be of importance in leading to the enrichment of rare metals in the deposits. Small amounts of gold, silver and other valuable metals have been found in mid-ocean ridge sulphide deposits, and more recently in island arc varieties. Sulphides from the Mariana Volcanic Arc in the Western Pacific, for example, have been reported as being significantly enriched in gold. Rare metal enrichments in island arc sulphides as compared with mid-ocean ridge varieties would be in accord with the more variable composition of the island arc rocks available for leaching than those on mid-ocean ridges.

The earliest reported, and still the most economically viable, polymetallic sulphide deposits occur in the Red Sea as a sulphide rich mud in the Atlantis II and some other deeps. These deposits have been the subject of a concerted technological and economic feasibility study, and are likely to be mined when economic conditions improve.

Following the discovery of the Red Sea deposits, similar occurrences were sought, and found, on mid-ocean ridges, largely using an exploration model developed on the basis of an understanding of the Red Sea deposits. Some of these are in EEZ's, such as deposits on the East Pacific Rise and Juan de Fuca Ridge off western North America, but most are not. By contrast, more recently discovered deposits in island arcs and back-arc basins are largely within EEZ's, and are the focus of considerable attention at the present time.

The S.W. Pacific back-arc basins from the Manus Basin to the Lau Basin have been the subject of several investigations for seafloor hydrothermal deposits. In the Bismarck Sea, several hydrothermal deposits have been found on or near the plate boundary, such as sulphides in the centre of the Manus Basin, and Mn oxides both in association with the sulphides and near Manus Island. Hydrothermal gas anomalies in the water column in the Manus Basin have confirmed the contemporary nature of the hydrothermal activity. Similar deposits, including iron silicates, have been found in the Woodlark Basin. However, it is in the Lau and North Fiji Basins that the major S.W. Pacific hydrothermal deposits found to date are located. These include sulphide, silicate and oxide deposits of considerable extent. Wide

haloes of hydrothermally enriched sediments surrounding the deposits
and widespread methane anomalies in the water column attest to the
continuing activity of the hydrothermal systems that formed them.

In contrast to the back-arc basins, S.W. Pacific Island arcs have
not been found to host such a rich suite of hydrothermal deposits, only
low temperature Mn oxides having been found there. Higher grade
deposits may occur below the seafloor and/or in craters in the
immediate vicinity of the vents.

Technological needs for finding and assessing seafloor
hydrothermal deposits include better optical imaging systems and water
column sensors for exploration, and in-situ samplers such as drills and
corers for proving the extent of the deposits.

4.   MANGANESE NODULES

Manganese nodules have been long regarded as possible future ores of
Ni, Cu and Co, principally of Ni and Cu. The best Ni and Cu rich
varieties occur in the North-eastern tropical Pacific and in the
central Indian Ocean, both areas the subject of mine site claims under
the Pioneer Investor provision of the Law of the Sea Convention. More
recently, Ni and Cu rich nodules have been found in the EEZ's of some
of the south equatorial Pacific island nations such as Kiribati, but in
general the abundancies of the nodules are low and thus metal
quantities per unit area of seafloor are also low.

Nickel and copper enrichment in the south equatorial Pacific EEZ
nodules appears to be related to organic fluxing of metals under the
equatorial zone of high biological productivity. Highest Ni and Cu
contents occur in the depth range of the calcium carbonate compensation
depth which occurs between about 4900-5300 m in the south equatorial
Pacific. At depths both above and below this, the Ni and Cu content of
the nodules decreases. This can be explained by the concentration of
metal bearing organic material at the CCD and its decay there
liberating the metals for uptake by the forming nodules. Above the CCD
the organic material is diluted by $CaCO_3$, inhibiting its role as a
diagenetic supplier of metals to nodules, and where the seafloor is
deeper than the CCD much decay of organic material probably takes place
in the water column below the CCD, thereby reducing the amount
sedimented and able to mediate in diagenetic reactions leading to
nodule formation on the seafloor.

Of considerable recent interest in South Pacific EEZ's is the
discovery there of large deposits of Co-rich nodules (Fig. 1). These
occur south of the equatorial zone of high biological productivity and
organic processes are thought to play little part in their Co
enrichment. The great abundance of these deposits, often more than 30
kg/m² and containing up to 0.6% Co, is probably related to their
occurrence under areas of Antarctic Bottom Water flow, the erosive
nature of which drastically reduces sedimentation rates which in turn
leads to high nodule abundances. Cobalt enrichment may result from
slow growth of the deposits.

Resource assessments for nodules have been carried out in the
EEZ's of some of the British Commonwealth CCOP/SOPAC nations in the

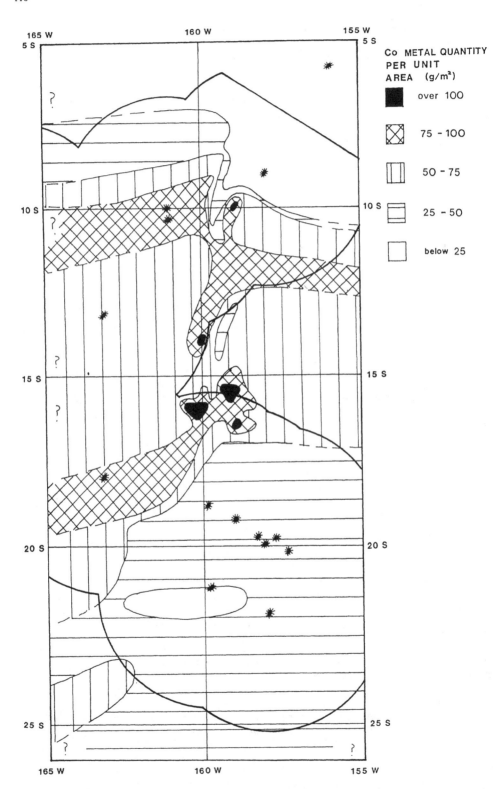

Figure 1

South Pacific, with surprisingly positive results. In the Cook Islands
EEZ, for example, the Ni equivalent metal quantities (Ni+Cu+Co) in all
nodule deposits present in quantities of more than 5 kg/m² (a commonly
proposed cut off abundance for mining purposes) are in excess of
100 m.t. In the much smaller Phoenix Islands EEZ (Republic of
Kiribati), Ni equivalent metal quantities are around 12 m.t. These are
estimates of in place quantities only and as such do not indicate
either potential recoverability or mineable amounts.

5.   COBALT-RICH CRUSTS

Cobalt bearing Mn crusts have been known for almost as long as Mn
nodules, but have only recently been thought of as a potentially
economic deposit. They coat exposed rock substrates preferentially
between 800-2400 m depth, and contain up to about 1.5-2% Co (average
about 0.9%). They are thought to receive their metals by diffusion
from the oxygen minimum zone which has been found to contain both
elevated Mn and Co contents. Unlike Mn nodules these are almost
exclusively EEZ deposits, but in some EEZ's collectively contain less
valuable metals than do the nodules (see below).
        Resource assessment studies for Co-rich crusts have been carried
out in the EEZ's of both US Trust Territories in the Western Pacific
and some British Commonwealth CCOP/SOPAC island nations in the South
Pacific. Approximate amounts of metals, for example, in Cook Islands
EEZ crusts are 930,000 tonnes of Co and 630,000 tonnes of Ni, and in
Phoenix Island crusts 69,000 tonnes of Co and 56,000 tonnes of Ni.
There are also estimates of in place quantities only.

6.   RELATIVE NODULE/CRUST ASSESSMENTS

The Cook and Phoenix Islands EEZ's represents the only complete EEZ's
to date where nodule and crust assessments have been carried out in
concert. Interestingly, the figures given above demonstrate much
greater quantities of the potentially economic metals Ni, Cu and Co in
the nodules than in the crusts in these EEZ's, and indeed even greater
quantities of Co alone in the nodules than in the crusts.
        Unlike in the case of nodule mining, techniques for Co-crust
mining have yet to be worked out. Separation of the deposits from
attached substrate will be a major concern, and up to 50% dilution of
crust by substrate has been suggested as not an unreasonable
approximation of recovered ore. This would bring the Co values in
recovered crust material down to or below Co values in Co-rich nodules.
It has further been suggested that only one or two Co-rich crust mining
operations might be needed to satisfy future world demand for marine
Co.
        Considerations such as these, coupled with the much greater
amounts of Co in nodules than in crusts in the EEZ's studied could
suggest that Co-crust mining will not be a viable proposition and that
marine Co could better be obtained from EEZ nodules.

THE ASSESSMENT OF AGGREGATE RESOURCES FROM THE UK CONTINENTAL SHELF

D.A. ARDUS and D.J. HARRISON
British Geological Survey
Edinburgh
Scotland
U.K.

ABSTRACT. In the UK marine-dredged sand and gravel make a major
contribution to meeting the demand for raw materials for the
construction industry. In order to develop an improved awareness of
marine aggregate resources, and to assist in planning, the Department
of the Environment and the Crown Estate Commissioners (jointly
responsible for the management of the aggregate dredging licencing
system) have recently commissioned the British Geological Survey to
undertake a programme of research designed to investigate offshore
resources of sand and gravel.

A desk study to summarise available information and to locate
potential marine aggregate resources in the Southern North Sea
commenced in 1986. This review identified an area off East Anglia for
more detailed investigation and subsequently this was surveyed in
1986-87.

The research has established the distribution and potential
quality of marine sand and gravel resources. Geophysical, sampling and
coring techniques have allowed the thickness and lithology of the
Holocene sediments and of the underlying Pleistocene sediments to be
determined. The survey has shown the complexity of the surface and
sub-surface sediments and the fundamental importance of an
understanding of the regional geology of the area to the determination
of marine aggregate resource potential.

Further studies are planned in the remaining areas of the UK
continental shelf.

1. INTRODUCTION

The UK marine-aggregate industry has its roots in the dredging of
ballast for sailing ships in the 16th Century. By the 18th Century
dredging had developed into a major port activity and provided the
principal source of income to Trinity House, the organisation
responsible for the lighthouses, lightships and buoys around southern
Britain. However, it is only the past 25 years that the industry has
expanded rapidly to provide considerable supplies of aggregate for the
construction industry.

D. A. Ardus and M. A. Champ (eds.), Ocean Resources, Vol. I, 113–128.
© 1990 *Kluwer Academic Publishers. Printed in the Netherlands.*

National guidelines (Department of Environment, 1982) now
encourage the use of marine-dredged aggregates when this does not cause
unacceptable damage to sea fisheries or coastal erosion. Current major
uses of marine aggregate are for concrete and as bulk fill in highway
schemes, port construction works and land reclamation. Minor uses
include beach replenishment, stabilisation of offshore oil and gas
platforms and in sand filters at water treatment works.
      The sea bed and foreshore, between high and low watermarks, are
ancient Crown possessions. The Crown Estate own approximately 55% of
the foreshore, most of the bed of the Territorial Sea to the 12 mile
limit, and the rights to explore and exploit the natural resources of
the Continental Shelf, with the exception, as on land, of hydrocarbons.
In 1961 the Crown Estate Act established the Crown Estate Commissioners
(CEC) as a corporate body to manage lands held in right of Crown which,
since 1760, have passed annual revenues, the Consolidated Fund, to
Parliament. This presently amounts to some £30 million (m) annually,
of which £1.5 - £1.75 m is derived from offshore dredging. The Crown
Estate is not a government department and management and licencing
arrangements depend on their common law powers as owners of resources.
The only statutory consent required outside port areas relates to
navigational interference and is administered by the Department of
Transport under the Coastal Protection Act of 1949.
      In 1964 the UK Continental Shelf Act confirmed the right of the
Crown Estate Commissioners to grant enforceable concessions for marine
sand and gravel extraction. Since then marine aggregates have grown in
importance and now represent some 16% of the total annual requirement
for sand and gravel in England and Wales. The market is predominantly
in London and the south-east of England where one-third of the total
offshore production meets some 30% of the market demand. In some
counties in the south of England marine dredging provides 35-40% of the
total material used. Over the last few years tonnage landed has
increased annually in response to planning policies, which constrain
the extraction of potential gravel resources inland, and increased
market penetration. Additionally, remaining reserves onshore are often
located in prime agricultural land or where environmental
considerations present a major limiting factor (Uren, 1988).
      Since 1970 the increase in dredging has not been matched by the
rate of release of new licenced ground and the lack of identified
resources has been a developing concern.

2.   PRODUCTION

Marine-dredged sand and gravel production in the UK has risen from 7 x
10^6 tonnes per annum in 1965 to 16 x 10^6 tonnes per annum in 1987.
This represents a significant contribution to a total national
requirement for aggregates of over 100 x 10^6 tonnes per annum. A
further 3 x 10^6 tonnes of marine material were exported to France,
Belgium, Holland and West Germany in that year. This compares with
production for 1983 of 57 x 10^6 tonnes landed in Japan, 9.5 x 10^6
tonnes in Denmark and 7.6 x 10^6 tonnes for Belgium, Holland and West
Germany combined.

Currently 11 companies operate a total of more than 50 ships with a gross hopper capacity of some 50,000 m³. The majority of ships are trailer suction dredgers varying in size from the largest at 5487 gross registered tons (GRT) with a cargo capacity of 7500 tonnes, to 4 in the 3000-4000 GRT range, each with a capacity of c. 4500 tonnes, and 22 of 1000-3000 GRT, with the remainder less than 1000 GRT.

The vessels operate in the following areas of the UK coast (Fig. 1): East Coast, South Coast, Thames Estuary, Bristol Channel, Humber and Liverpool Bay, the first mentioned being the major area both in terms of production and proximity to the major market (Fig. 2).

Dredging is restricted mainly to <35 m water depth, although newer vessels have the capacity to dredge to 45 m using submersible dredge pumps located mid-way in the dredge pipe.

3.   LICENCING

The procedure established by CEC requires applicants to demonstrate that they possess the capability to meet licence conditions (Parrish, 1988). Licences take two forms, one for prospecting the other for production. The prospecting licence is issued for 2 or 4 years, is renewable, and allows side-scan sonar, seismic profiling, grab sampling and coring, together with dredger sampling limited to 1000 tonnes per two years. Dredger sampling was previously the standard prospecting procedure and remains an important assessment of commercial value and viability. Although prospecting licences are commercially confidential, GEC consults with the Ministry of Agriculture, Fisheries and Food (MAFF) concerning fisheries interests prior to their issue.

Following successful prospecting, applications for production licences follow an informal consultation procedure whereby CEC have agreed not to issue a licence if any substantive objection raised by a relevant government department cannot be resolved.

Assessment of production licences proceeds initially with Hydraulics Research Ltd. (H.R. Ltd., formerly the Hydraulic Research Station of the Department of Environment) regarding potential damage to any adjacent coast. Data provided to H.R. Ltd. are the proposed working area, the prospecting report and the forecast maximum annual extraction. H.R. Ltd. appraise distance offshore and depth of water to ensure that no draw-down of beach material or change in wave-refraction pattern will occur which might alter longshore transport and shore stability and consider the effect on protection afforded by offshore bars to avoid consequent coastal erosion (Brampton, 1985).

Companies are required by CEC to meet the cost of any necessary desk studies, computer simulations or field research recommended by H.R. Ltd. should they wish to proceed. Such studies, which necessarily take a worst case view to ensure that the coastline will suffer no detrimental effects, may result in approval, revision, or rejection of proposals.

Given H.R. Ltd. clearance, and an absence of obstacles such as pipelines, power and telephone cables, proposals are referred, together with the H.R. Ltd. report and the CEC view to the Minerals Division of the Department of Environment for a Government View. This is obtained

by a process of consultation with other relevant divisions of DOE, MAFF
(concerning fisheries and coastal protection), Department of Transport,
Ministry of Defence and Department of Energy. Departments consulted as
necessary include the Welsh Office, Scottish Development Department and
the Department of Industry. Other bodies consulted at this stage are
the local coast protection authority, the mineral planning authority
and the regional water authority. These consultees in turn seek
observations from other bodies whose interests may be affected by the
proposals, e.g. Nature Conservancy Council, Trinity House, etc.
     Resolution of any substansive objections are pursued by
negotiation, further research or by changing the area or licence
conditions. Licences are not issued when objections cannot be resolved
as, for example, should the Government View reflect a MAFF belief that
the impact of dredging on commercially important fish stocks and their
food chain will be significant.
     Production applications are commonly for areas up to 50 km^2 and
sometimes for areas less than 20 km^2. The actual area being dredged at
any one time is small. Altogether, assuming an average dredging depth
of 1 m, the total area of seabed directly affected annually is
approximately 9 km^2.
     Production licences define the permitted working area, maximum
annual recovery and specific conditions. The latter may, for example,
stipulate a maximum permitted dredging depth below seabed and/or
require a layer of sandy gravel to be left as a covering cap.
Alternatively, time or seasonal constraints may be imposed. Production
licences may be terminated by either party at six months notice.
     Dredging is monitored through audited records and returns, by
side-scan sonar surveys which reveal evidence of workings and by
observation from fishery patrol flights. Unauthorized activity is
curtailed by the sanction of temporary or permanent withdrawal of the
licence.
     A Code of Practice for the Extraction of Marine Aggregates was
established in 1981 between the Marine Section of the Sand and Gravel
Association and the Association of Sea Fisheries Committees to promote
mutual cooperation and minimise potential interference to the
activities or damage to the resources of either interest group.
(Appendix 4, Nunny and Chillingworth, 1986).
     Presently CEC, DOE and the dredging companies are reviewing the
Government View and licence procedures while CEC and DOE are
establishing the criteria applicable to applications in sensitive
situations, which will define when an environmental impact assessment
is required under the European Community Directive on Environmental
Assessment.
     The review aims to take account of the concerns of industry,
environmental groups and the public who presently may be unaware of
proposals because no requirement exists to publicise applications.

4.   REQUIREMENTS

A study of Marine Dredging for Sand and Gravel undertaken for DOE
(Nunny and Chillingworth, 1986) recommended amongst other things that
there was a lack of national awareness of the location and quality of

marine resources beyond established areas of reserves and that the considerable existing information concerning the UK continental shelf should be coordinated and supplemented where necessary by further field studies, to provide the regional knowledge of resource potential. This, together with efficient exploitation of existing reserves through avoidance of partial dredging and over-sanding (dumping of excess fines) and better prospecting procedures and reserve management, addressed the immediate concerns regarding continuity of reserve availability.

Further recommendations included the proposal that reasons for licence refusals should be categorized, that sufficient data on reserves and environmental considerations be submitted with each licence application and that forward planning of reserve releases should allow time for broad environmental aspects to be considered prior to licence application.

Research into potential development of new techniques for prospecting and reserve management were also identified in geophysical and geochemical techniques, deposit coring and in-pipe aggregate monitoring.

Coincident with the first recommendation, a series of desk studies were commissioned by CEC from British Geological Survey (BGS). These are complemented by subsequent surveys by BGS the first funded by DOE and CEC, and thereafter, by DOE, CEC and BGS to establish a national awareness of aggregate resources in order that a rational licencing and resource management policy may be followed.

5.   PREVIOUS WORK

BGS began systematic geological survey of the UK continental shelf (UKCS) in 1966 and completed the reconnaissance data acquisition for the inner shelf area within 200 m water depth, together with the slope beyond into the Rockall Trough and the Faeroe-Shetland Channel to 2000 m depth, in 1986. The extended designated area to the west remains to be comprehensively surveyed. Assimilation of available data from hydrographic sources, from commercial exploration and site investigation (particularly for oil and gas) and from academic studies has complemented the geophysical surveys, sampling and drilling carried out by BGS (Ardus, 1985).

In the UKCS survey simultaneous gravimeter, magnetometer and air-gun or water-gun or sparker analogue seismic have been run, together with boomer or pinger high-resolution profiling and sidescan sonar (Dobinson and others, 1982). These data have been calibrated by grab sampling, remotely operated sediment and rock coring, to 6 m penetration below seabed (Pheasant, 1984), and by shallow drilling (Ardus and others, 1982), the latter constrained to a maximum penetration of 300 m below seabed by the absence of blow-out prevention equipment on the site-investigation drilling vessels used.

Over 225,000 km of multiple seismic instrument traverses have been run, samples and cores obtained from more than 30,000 stations and over 500 boreholes have been drilled. The common geophysical traverse interval is of the order of 10 km and this grid is complemented by

commissioned digital seismic surveys in some areas. Underway, seafloor
gamma-ray spectrometry has also been run in selected areas.
     Data are held in a comprehensive computerised data base and
interpretations are being published in a series of maps compiled at
scales of 1:250 000 and 1:1000 000 showing gravity and aeromagnetic
anomalies, solid geology, Quaternary geology and seabed sediments. A
series of regional reports will complement the maps. About 73% of the
maps are published and complete sets of maps and reports are scheduled
for completion by early 1993.
     Concurrent with these studies, which have been funded largely by
Department of Energy (DEn) during the last decade, a separate,
confidential assessment of commercial hydrocarbon exploration data has
been performed for DEn. However, the Regional Mapping Programme does
not provide sufficient detail to allow an adequate offshore aggregate
resource assessment to be made and the work now being undertaken for
DoE and CEC, with respect to sand and gravel resources, is analogous to
this hydrocarbon assessment and pursues parallel objectives.
     The aim of identifying national resources will not preclude the
necessity for companies to undertake their own evaluation of potential
deposits in order to establish the extent and value of their reserves.

6.   GEOLOGICAL BACKGROUND

The initial CEC desk-study area was the Southern North Sea south of
53° N, from the English coast to the median line between UK, France,
Belgium and Holland.
     The North Sea basin, subjected to tensional forces, has been a
major centre of deposition since the Permo-Trias, with faulted grabens
giving way in the late-Cretaceous and Tertiary to basin-wide subsidence
which continues today. In the Southern North Sea a major unconformity
occurs chronologically close to the base of the Quaternary, as defined
in the Netherlands and eastern England. This represents the change in
conditions which occurred some 2.47 Ma BP (McDougall, 1979) when a
major delta began to build out across what is now the Netherlands and
the Southern North Sea (Cameron and others, 1987). To the North marine
conditions prevailed and Quaternary sediments are largely conformable
on Pliocene sediments and onlap westward onto older rocks toward the
Scottish coast.
     The deltaic sequence reflects alternating marine transgressions
and regressions coincident with temperate and cooler climatic periods
while to the north more continuous marine sedimentation continued
(Stoker and others, 1983, 1985; Stoker and Bent, 1987). In the Upper
Pleistocene major glaciations occurred in the Elsterian, Saalian and
Weichselian. Evidence for these is drawn from the presence of major
unconformities into which large valleys, eroded into the underlying
sediments, are incised. In these valleys the initial infill of ill-
sorted sediment is overlain by glacio-marine or glacio-lacustrine
sediments which pass upward into marine interglacial deposits. Similar
open valleys exist in the present seabed containing significantly more
Holocene deposits than occur on the surrounding seafloor.

The gravels are mainly composed of flint, but other rock-types, particularly quartzite and phosphorite, are locally important and gravel-sized shell debris and large shells may in some areas constitute 30% or more of the gravel fraction. However, over most of the Southern North Sea area, flint composes over 90% of the gravel and the character of the flint varies depending on its occurrence and geological history. The flint material has been derived directly from Cretaceous outcrops, mainly the Chalk, or has been reworked from Tertiary and Quaternary deposits. Quartzose gravel may be derived from a number of sources of which fluvial erosion of the Triassic rocks of the English Midlands may be the most important. Phosphorite in the gravels is derived from the erosion of a conglomeratic layer at the base of the Red Crag formation.

The present distribution of gravel deposits is the result of marine reworking of former fluvial accumulations during the early Holocene. They are relict deposits, now only reworked to a minor degree as indicated by overgrowths and encrustations of marine organisms.

7. AGGREGATE STUDIES AND SURVEYS

Despite the developing understanding of Quaternary sediments and the increasing importance of marine aggregates, information on the distribution, quality and quantity of sand and gravel resources outside established licenced areas is sparse.

In 1986 the Crown Estate Commissioners engaged BGS to undertake a review of all available data in the Southern North Sea with the aim of summarising potential marine aggregate resources. This desk study drew together information from the BGS offshore survey programme and from commercial and academic sources.

Information concerning the bathymetry, tidal current regime, distribution and transport of seabed sediments, Quaternary and solid geology were reviewed in the context of the dredging industry to show the position of potential marine aggregate resources in the Southern North Sea area. The report (Balson and Harrison, 1986) included seven maps at 1:250 000 scale showing:

1. Bathymetry
2. Bedforms
3. Sample Points
4. Seabed Sediments and underlying geology
5. Surface distribution of coarse aggregate (gravel)
6. Surface distribution of fine aggregate (sand)
7. Licenced dredging areas

This desk-study was the first in a series of studies which will eventually cover all offshore areas around the British Isles. The second and third phases of this research programme were undertaken in 1988, covering the South Coast area of England and the Bristol Channel. A study of the East Coast area is due to start shortly.

An important objective of the desk-study programme is to identify areas suitable for more detailed resource assessment surveys using geophysical, sampling and coring techniques. The Southern North Sea study recommend an area off Great Yarmouth, East Anglia for further investigation and in 1986-87 the Department of the Environment and the Crown Estate Commissioners commissioned BGS to investigate the marine aggregate resources in this area (Fig. 3).

Data were obtained during a geophysical survey in October 1986 and a bottom sampling and coring survey in July 1987. The geophysical survey was based on a regular 5 km x 5 km grid of traverses with closer spaced (1.5 km) lines over areas of likely sand and gravel resources identified in the desk-study. Geophysical techniques used were high-resolution surface and sub-tow boomer seismic reflection profiling and sidescan sonar. Decca Main Chain was used for position fixing and provided adequate accuracy. Subsequent surveys, however, will use a higher level of navigational control.

Shallow cores and grab samples provided the necessary data to calibrate the geophysical interpretation and material for laboratory testing. Seabed samples were obtained using a clamshell grab of 0.65 m^3 capacity (Plate 1). Cores were obtained using a standard BGS vibrocorer fitted with a 6 metre barrel of 4 inch or 6 inch diameter (Plates 2 & 3). Additional seabed data were obtained using an underwater camera system.

The grading characteristics of each bulk sample and selected core samples were determined in the laboratory in order to obtain relative proportions of gravel, sand and fines in the sediment and to classify the marine aggregate resources. The carbonate contents of both gravel and sand fractions were also determined. Detailed lithological descriptions of borehole cores and grab samples were prepared and many cores were subsampled for palaeontological analysis.

The results of the survey are included in a comprehensive report (Harrison, 1988) and are summarised on eight accompanying 1:100 000 scale maps which illustrate the geology and resource potential of the area.

Sediments of Pleistocene age underlie most of the seabed in the survey area and are overlain by a cover of Holocene (superficial) sediment. Sand deposits are extensive, occurring in both superficial sediments and in the underlying Pleistocene strata; gravelly sediments are mainly restricted to the superficial sediment. The Pleistocene has been divided into seven formations based on their seismic character, lithology and stratigraphical relationships (Fig. 4, Table 1). They represent a complex sequence of marine and non-marine deposits mostly of sands, shelly sands, silts and clays. Gravelly sediment has been recorded from several parts of the succession, but amounts are relatively insignificant except for one part of the sequence which locally contains gravel layers.

In the south of the area early Tertiary clays of the London Clay Formation lie at, or near, seabed and form a slightly undulating surface dipping gently to the north-east where overlying Pleistocene sediments are over 100 m thick.

Plio-Pleistocene (Red Crag Formation) shallow-water, marine, shelly sands infill hollows in the London Clay and are overlain by early Pleistocene clays and muds of the Westkapelle Ground Formation deposited in a marine open-shelf environment. The overlying succession of Lower and Middle Pleistocene age reflects the massive seaward extension of the delta system of North-West Europe into the North Sea Basin. These deltaic sediments are predominantly fine- or medium-grained sands, although gravel and clay layers are present in the Yarmouth Roads Formation, the youngest pre-glacial sediments.

A late-Middle Pleistocene regional unconformity associated with three major glaciations, the Wolstonian (Elsterian), the Anglian (Saalian) and the Devensian (Weichselian), separates these predominantly marine sediments from the shallow marine sands and clays of the Eem Formation and the brackish-marine silty clays of the Brown Bank Formation which were deposited during the marine regression at the onset of the Devensian, the most recent glaciation. Thereafter, sea level continued to fall and the area was exposed and subject to periglacial processes in the late Devensian.

In the early Holocene the area was drained by river systems and the fluvial deposits, together with the underlying Pleistocene sediments, were reworked in the transgression associated with the earliest marine incursion more than 9000 years BP. Marine conditions were established after c. 7000 years BP and present sea level some 5000 years BP when the modern tidal regime was established.

Beach deposits and near-shore deposits were modified following the transgression to form sheet sands from which tide and wave action have removed the fine sand fraction.

The seabed is particularly rich in gravel in two large areas off the East Anglian coast. In these areas, the surface layers typically contain up to 60% flint gravel and small amounts of fines, although the sub-surface layers (below 25-30 cm) contain a larger proportion of sand and typically 15-30% gravel. The gravelly sediments occur mostly in water depths of less than 35 m and form a veneer less than 1 m in thickness and locally less than 0.1 m thick. Thicker accumulations occur as the dredged remnants of former river channels and terraces.

8.   CONCLUSIONS

The research programme has clearly identified the distribution and potential quality of marine sand and gravel resources in the survey area. Although several areas off East Anglia are dredged for marine aggregates this survey indicates the occurrence of other potential resource areas lying outside those currently licenced for dredging. The geology of the surface and sub-surface sediments has been shown to be complex and the fundamental importance of a thorough understanding of the regional geology in the appraisal of marine aggregates has been demonstrated. This approach is being applied in other areas around the UK and is relevant and applicable elsewhere.

9.   ACKNOWLEDGEMENTS

This paper is published with the permission of the Director, British Geological Survey, National Environment Research Council and the approval of the UK Department of Environment.

10. FIGURES

Fig. 1 Licenced dredging areas

Fig. 2 Production of marine aggregate from principal sea areas (Source: Nunny and Chillingworth, 1986)

Fig. 3 Location of the survey area

Fig. 4 Schematic relationship between the stratigraphic units

11. TABLE

Table 1 Geological sequence

12. PLATES

Plate 1 Sediment sample obtained by hydraulically-powered clamshell grab

Plate 2 Vibrocore. 2.4 m of Holocene gravelly sand underlain by Pleistocene clay

Plate 3 BGS vibrocorer

13. REFERENCES

Ardus, D.A. (1985) 'Development of techniques for marine geological survey`, in P.G. Teleki (ed.), Marine Minerals Resource Assessment Strategies, NATO Advances in Science and Technology Series, Reidel Publishing Co., pp.81-97.

Ardus, D.A., Skinner, A.C., Owens, R. and Pheasant, J. (1982) 'Improved coring techniques and offshore laboratory procedures in sampling and shallow drilling`, Proc. Oceanology International 1982, Brighton, UK, 18pp.

Balson, P.S. and Harrison, D.J. (1986) 'Marine Aggregate Survey, Phase 1: Southern North Sea`, BGS Marine Report No. 86/38.

Brampton, A. (1985) 'Total effects of dredging on the coast. Problems associated with the coastline`, Proc. Conf. Newport, Isle of Wight, IOWCC, 11pp.

Cameron, T.D.J., Stoker, M.S. and Long, D. (1987), 'The history of Quaternary sedimentation in the UK sector of the North Sea Basin`, Jour. Geol. Soc., London, Vol. 144, pp.43-58.

Department of Environment (1982) 'Guidelines for aggregate
    provision in England and Wales', DOE Circular 21/82.

Dobinson, A., Roberts, P.R. and Williamson, I.R. (1982) 'The
    development of a control system for simultaneous operation of
    seismic profiling equipment', Proc. Oceanology International,
    1982, Brighton, UK, 16pp.

Harrison, D.J. (1988) 'The marine sand and gravel resources off
    Great Yarmouth and Southwold, East Anglia', BGS Technical Report
    WB/88/9/C Marine Geology Series.

McDougall, I. (1979) 'The present status of the geomagnetic
    polarity time scale' in McElhinny, M.W. (ed.), The Earth,
    Its Origin, Structure and Evolution, Academic Press, London,
    pp.543-566.

Nunny, R.S. and Chillingworth, P.C.H. (1986) 'Marine dredging for
    sand and gravel', Department of the Environment, Minerals
    Division, HMSO, London.

Parrish, F.G. (1988) 'Management of the UK marine aggregate
    dredging', Jour. Underwater Technology, Vol. 14, No. 2,
    pp.20-25.

Pheasant, J. (1984) 'Microprocessor controlled seabed rockdrill/
    vibrocorer', Jour. Underwater Technology, Vol. 10, No. 1,
    pp.10-14.

Stoker, M.S., Skinner, A.C., Fyfe, J.A. and Long, D. (1983)
    'Palaeomagnetic evidence for early Pleistocene in the central
    and northern North Sea', Nature, Vol. 304, No. 5924, pp.332-334.

Stoker, M.S., Long, D. and Fyfe, J.A. (1985) 'The Quaternary
    succession in the central North Sea', Newsletters on
    Stratigraphy, Vol. 14, pp.119-128.

Stoker, M.S. and Bent, A.J.A. (1987) 'Lower Pleistocene deltaic
    and marine sediments in boreholes from the central North Sea',
    Jour. Quaternary Science, Vol. 2, pp.87-96.

Uren, M.J. (1988) 'The marine sand and gravel dredging industry of
    the United Kingdom', Marine Mining, Vol. 7, pp.69-88.

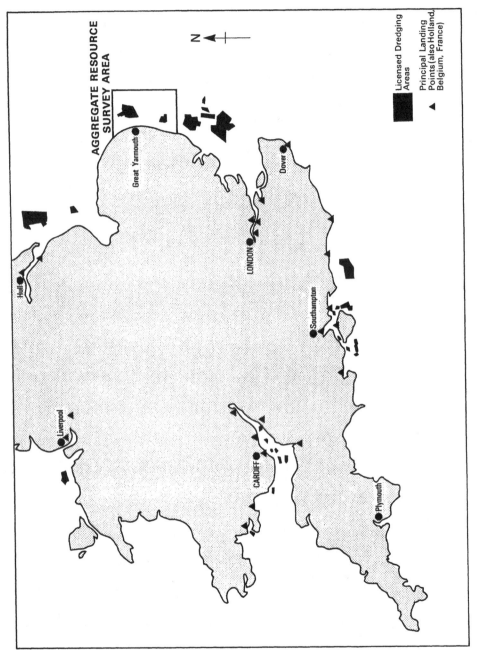

Figure 1. Licensed Dredging Areas

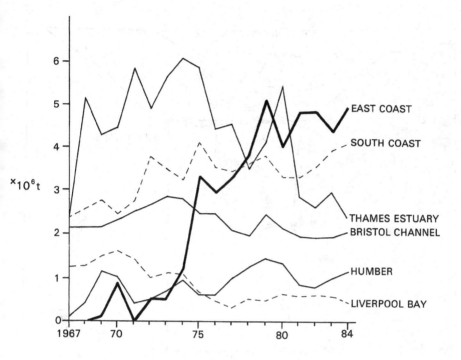

Figure 2   Production of marine aggregates from principal sea areas
(Source: Nunny and Chillingworth 1986).

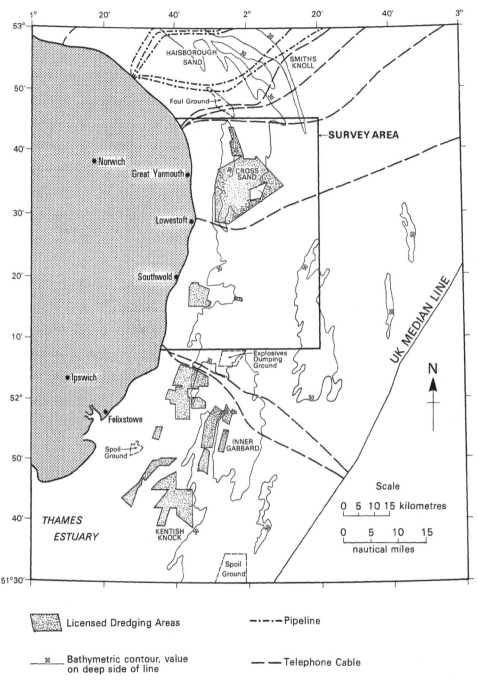

Figure 3 Location of the survey area

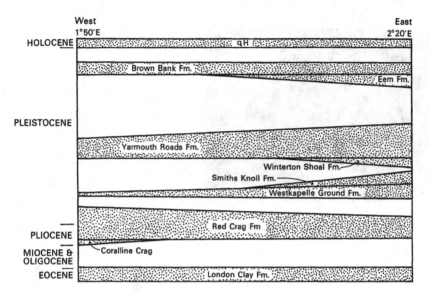

Figure 4   Schematic relationship between the stratigraphic units

QUATERNARY

HOLOCENE - superficial sediments (sands and gravelly sands). Up to 20m thick.

PLEISTOCENE

Upper Pleistocene

Brown Bank Formation - silty clay.   30m maximum thickness.
Eem Formation   - muddy sand. 20m maximum thickness.

unconformity

Lower Pleistocene
Yarmouth Roads Formation   - sand with clay and gravel layers. Up to 70m thick.
Winterton Shoal Formation   - sand. 10m thick.
Smith's Knoll Formation   - sand. 20m thick.
Westkapelle Ground Formation - clay and muddy sand.   50m maximum thickness.

PLIOCENE (TERTIARY) to early PLEISTOCENE

Red Crag Formation - shelly sand. 60m maximum thickness.

unconformity

TERTIARY

PLIOCENE

Coralline Crag Formation - shelly sand. 10m maximum thickness.

EOCENE

London Clay Formation - stiff clay. > 50m thick.

Table 1   Geological sequence

Plate 1. Sediment sample obtained by hydraulically-powered clamshell grab

Plate 2. BGS vibrocorer

Plate 3. Vibrocore.
2.4m of Holocene gravelly sand
underlain by Pleistocene clay

EXPLOITATION OF MARINE AGGREGATES AND CALCAREOUS SANDS

ALAIN PHILIPPE CRESSARD *
CLAUDE AUGRIS **

ABSTRACT. Important quantities of sand and gravel occur on the French
continental shelf, particularly in the Channel and on the Atlantic
Coast.
     Production has been increasing in the last ten years, however,
only by very small companies using rather small dredgers. Mining
licences are issued by the Ministry of Industry for prospecting,
research and exploitation of marine aggregates.
     Mining licences and extraction permits require a unique case
including an evaluation of the consequences of dredging on the marine
environment.

1.   RESOURCES AND RESERVES OF THE FRENCH CONTINENTAL SHELF

With its 4000 km of coast and the extension of its area of economic
interest to 200 nautical miles offshore, France presents a wide
maritime facade. Legislation concerning submerged minerals was drawn
up in 1968. Among these materials, there is an important quantity of
sand and gravel on the French continental shelf, particularly in the
channel and on the Atlantic coast.
     These sand and gravel (siliceous and calcareous) deposits formed
at variable rates, and include clay beds full of fragmented tests of
marine organisms. These represent two distinct kinds of accumulation
dependant on continental or marine sedimentary processes.
     Alluvial deposits of fluvial origin have been formed during the
marine regressions of the Quaternary ice-age, when, with water bound up
in ice sheets on the continents, the continental shelves were sub-
aerialy exposed. Consequently, river flow lengthened and erosion
sought to establish new equilibrium profiles. This phenomenon has
allowed the accumulation of important deposits in valleys which are now
submerged by the subsequent marine transgression (Fig. 1).

*   IFREMER-DERO/PG, 66 Avenue d`Iena, 75116 Paris, France.
**   IFREMER - DERO/GM, BP 70, 29263 Plouzane, France.

D. A. Ardus and M. A. Champ (eds.), Ocean Resources, Vol. I, 129–137.
© 1990 *Kluwer Academic Publishers. Printed in the Netherlands.*

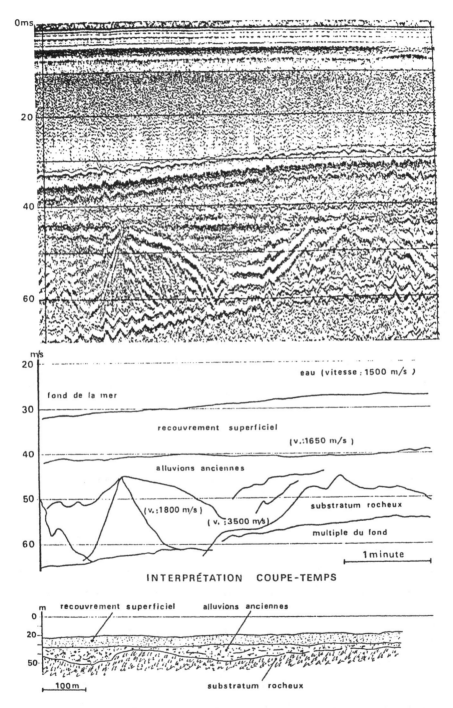

Figure 1. Seismic recording and interpretation of a palaeovalley

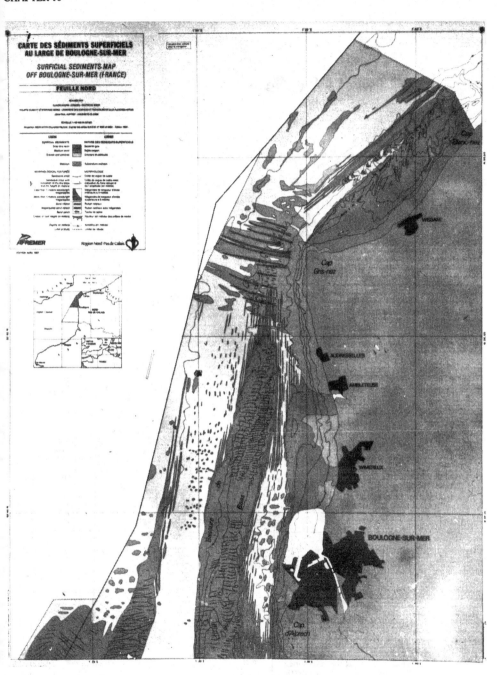

Fig. 2 - Superficial sediments map of Boulogne sur Mer

Ressources possibles	Ressources reconnues et localisation (Mm³)		Réserves (Mm³)
Côte de la Manche centrale et orientale On trouve d'importantes paléovallées dans toute la Manche, plus d'une centaine de Gm³ de ressources possibles	Wissant Boulogne Lobourg Vergoyer Dieppe Baie de Seine	10 60 1 330 225 970 3 000	10 — — — 100 400
Côtes du Massif armoricain Ressources importantes au-delà de 50 m de profondeur, plusieurs dizaines de Gm³ de ressources possibles	Saint-Malo Saint-Brieuc Nord Finistère Ouest Finistère Lorient Loire	450 2 635 387 875 2 530 15 630	— — — — 18 24
Côtes de l'Aquitaine Il reste la région de la Gironde et de l'Adour à prospecter, de l'ordre de la centaine de Gm³ de ressources possibles	Pertuis Breton Pertuis d'Antioche	1 100 1 800	30 35
Côtes de la Méditerranée Il reste à prospecter le golfe du Lion (plusieurs dizaines de Gm³) et la Corse (sans doute ressources inférieures au Gm³)	Entre Marseille et la frontière italienne	150	—
Pour la France, ressources possibles très importantes, centaines de Gm³		≈ 30 000	≈ 600

Figure 3.   Areas where prospecting has been undertaken.

At the present time, terrigenous detritus carried from rivers onto the shelves is mainly silt and fine sand. In this sand, we find an important fauna which produces calcareous particles which distort the grain size of the deposit and increase the carbonate percentage.

Marine sediments form the superficial layer on the continental shelf and are underlain in palaeovalleys by continental alluvium. These deposits form by the reworking and redistribution of deep-seated sediments by currents and may develop significant thicknesses to create dunes. Some of these have a high calcareous content consisting of shell fragments. The possibility of utilizing those sediments for concrete has been considered (Fig. 2).

Ifremer has undertaken the investigation and the evaluation of usable aggregates resources on the French continental shelf. An inventory of resources, including all accumulations located in less than 50 m water depth at low spring tide, identifies an estimated 30×10^9 m^3 of sand and gravel (siliceous and calcareous). Workable reserves are much less as several conditions must be met to make resources economically viable (Fig. 3). These include the following:

- the depth of water must allow dredging with available equipment (35 m of water);

- the exploitation must not disturb any human activity around the site (fishing, submarine cables, sea routes, Naval experimental areas);

- the location of the workable reserves must not be in an ecologically sensitive area.

- the quality of materials must allow their immediate utilization with a minimum of rejected tailings which could involve pollution and disturbance.

Presently in France, all these factors reduce the resources to 600×10^6 m^3 of workable reserves.

2.   DEVELOPMENT OF ACTIVITIES

The exploitation of marine sand and gravel is a very old activity and, by the seventeenth century, coastal communities were allowed to extract sand necessary to meet their needs. The increase in building activities and public works has caused a rise in aggregates production and, though most demand has been satisfied by opening of quarries, production of marine materials has clearly increased.

Thus, presently in France, the production of extracted sand and gravel from national waters, reaches 4.5×10^6 tons, equivalent to 1.5% of total output. This is low compared with other European countries, Great Britain for instance, and the proportion differs according to the regions, being quite low, for example, in the South of France (Figs. 4, 5 and 6).

DEPARTMENTS	PRODUCTION IN TONS	DEPARTMENTS	PRODUCTION IN TONS
Seine Maritime	750,000	Loire Atlantique	1,300,000
Ile et Vilaine	80,000	Vendée	160,000
Côtes du Nord	250,000	Charente Maritime	
Finistère	415,000	" "	700,000
Morbihan	290,000	Gironde	300,000
Approximate Total - 4,000,000			

Figure 4.   Sand and gravel production averages, 1980-1988

DEPARTMENTS	PRODUCTION IN TONS
Ile et Vilaine	150,000
Côtes du Nord	200,000
Finistère	150,000
Morbihan	50,000
Total	550,000

Figure 5.   Calcareous sands averages, 1980-1988

PORTS	IMPORTATIONS
Dunkerque	850,000
Calais	120,000
Boulogne	2,000
Fécamp	25,000
Caen-Ouistreham	40,000
Treguier	5,000
Roscoff	55,000
Brest	10,000
Total	1,017,000

Figure 6.   Sand and gravel import averages, 1980-1988

In contrast, marine extraction is more important in the west of the country with Brittany the major region producing 25% of the national production of marine aggregates.

Most vessels used for extraction of marine aggregates were equipped with grabs, but the need to maintain harbour channels, as well as to increase output for building activity and public works, has encouraged the development of more effective suction dredgers. Marine materials can now be discharged at prices competitive with those for aggregate derived from onshore quarries.

The arrival of bigger dredgers not only increased production but also the distances over which the materials could be conveyed economically. The exploitation of marine sand and gravel in France is effected by 20 companies all along the Channel and the Atlantic coasts, 50% working on a small scale (400 tons) and 50% using suction dredges of 1200 tons capacity. Brittany has 10 shipowning businesses which represents 50% of the French companies and 26% of the total tonnage. 34 vessels are equipped in France for the exploitation of marine aggregates, 20 are suction dredgers and 14 are fitted with grabs. All vessels presently operating effect three successive operations: extraction of the materials, transport and quayside discharge.

28 harbours along the Channel and the Atlantic coasts are equipped for dredgers. Two main techniques are utilized for discharge of materials, grabs or reverse hydraulic flow.

Grabs are used for discharging dry materials, working either from the vessel, when they are also used for extraction, or from the pier. Hydraulic discharge involves the liquefaction of the dredged material and their reverse flow through the shipboard pump. This second technique, which is faster, needs a draining dock, consequently the discharged materials are not available for immediate dispatch.

3.   APPLICATION FOR AN EXPLOITATION LICENCE

The exploitation of marine siliceous sand and gravel is controlled by mining legislation.

For a long time, aggregates were considered as quarry or as mining products according to their situation inside or outside the national waters (within or beyond 12 nautical miles) and were subjected to onshore legislation which was completely inadequate.

The July 1976 law concerning prospecting, research and exploitation of submerged mineral substances in the French public domain, the June 1980 decree for the application of this law and the decree of March 1980 concerning mining licences provide legislation relevant to the exploitation of marine aggregates.

Whatever their situation, inside or outside national waters, marine sand and gravel are now considered as mining products and exploitation is subject to the granting of a mining licence. The application for a mining licence and for local authorization requires a unique case including evaluation of quantities to be extracted annually. Production is subject to taxation which is considered due for the exploitation of the marine public domain.

Licences for prospecting research and exploitation of marine aggregates are now the responsibility of the Ministry responsible for mines. The implementation of this procedure will permit a rational exploitation of submarine mineral resources and will also permit the development of the extraction of marine sand and gravel.

4.   ENVIRONMENTAL STUDY

An environmental impact survey must be supplied when making application for mining licences and local authorization.

This study may result from consultation of published documents (with references) or result from experimental measures. The following points must be considered.

1.   Definition of the characteristics of the present environment (initial condition)

(a)   Oceanographic environment

- wave and current regime, sea surface and seabed characteristics

- turbidity and the nature of substances in suspension

- bathymetry and seabed morphology (sidescan sonar)

(b)   Biological environment

Definition of flora and fauna populations, indicating their situation with regard to the proposed exploitation area, particularly with regard to shells and spawning-grounds.

(c)   Fishing environment

Description of fishing activities (harbours, boats and fished species).

2.   Evaluation of the eventual consequences of proposed operations on the environment, on the extraction site and the littoral by fine materials in suspension and by wave action.

3.   Specific reasons why the specification has been chosen: geological and dredging characteristics ....

4.   Indication of the precautions to be taken to avoid or reduce detrimental effects.

To assist in the preparation of this case the proposer may call on the experience of experimental exploitation carried out in France by IFREMER.

5.   REFERENCES

Augris, C. et Cressard, A.P. (1984) 'Les granulats marins`,
   Publication du Centre National pour l'Exploitation des Océans`,
   - Rapport Scientifique et Technique, No. 51, 90 p.

C.E.A. - L.C.H.F. (1976) 'Etudes des gravières marines:
   comportement physique des particules fines remises en
   suspension`.

Cabioch, L. et Gentil, F. (1975) 'Benthos de la Manch et effets
   des attractions d'agrégats sur l'environnement marin", Station
   Biologique de Roscoff.

Cressard, A.P. (1975) 'Influence de l'exploitation des sables et
   graviers sur le milieu marin`, Terra Aqua 8/9.

Cressard, A.P. et al (1975) 'Effets des extractions de sables et
   graviers sur l'environnement`, C.I.E.M., Montreal.

Cressard, A.P., Chevasu, G. et Le Bris, J. (1977)
   'L'identification géotechnique des matériaux dans le cadre d'une
   recherche de granulats marins`.

Cressard, A.P. et Lemaire, J.P. (1977) 'Les granulats siliceux et
   calcaires du littoral français`.

Cressard, A.P. et Augris, C. (1981) 'Granulats du plateau
   continental, exploitation et constraintes d'environnement`,
   Séminaire 'Gestion régionale des sédiments`, Corse.

Cressard, A.P. (1982) 'Les granulats marins`, Annales des Mines,
   p.81-82.

Cressard, A.P. et Augris, C. (1982) 'Study of coastal erosion with
   the extraction of materials on continental shelf`, 4ème congrès
   international de l`A.I.G.I., New Delhi.

Debyser, J. (1975) 'Les problèmes de l'environnement lies a
   l'exploitation des sables et graviers marins`, Note technique
   CNEXO.

I.S.T.P.M. (1980) 'Effets de l'exploitation des granulats marins
   sur les activites halieutiques`.

S.N.A.M. (1980) 'L'armement français des navires sabliers`.

MAPPING, EVALUATION AND EXPLOITATION OF RESOURCES AND CONDITIONS IN
DANISH DOMESTIC WATERS

JOHN TYCHSEN
The Marine Raw Materials Survey
The National Forest and Nature Agency
Copenhagen
DENMARK

ABSTRACT.   The sea is exploited more intensively today than ever
before. We fish, we extract sand, gravel and stones as well as oil and
natural gas. We pump out sewage and dump rubbish. We dredge docks and
dump the often highly polluted sediment somewhere else on the seabed.
We use the sea for our recreation, for bathing, diving, sailing,
hunting and fishing. We can go on doing so, but we must exploit the
sea better so that all interests can benefit, and especially, we must
protect the cultural and natural assets of the sea.
    In this paper I shall describe the mapping and evaluation of
Danish EEZ resources, the balance between exploiting and protecting the
seabed and our experiences with monitoring the effect the extraction of
raw materials has on the seabed.

1.   ADMINISTRATION OF DANISH DOMESTIC WATERS

In Denmark the National Forest and Nature Agency, which is under the
Ministry of the Environment, administers the extraction of raw
materials and the protection of natural and cultural assets in Danish
domestic waters.
    The administrative basis is contained in the Raw Materials Act,
the Conservation of Nature Act and some international conventions,
among them the Ramsar Convention and the EEC Directive on Conservation
of Wild Birds.
    Pursuant to the Raw Materials Act, extraction permission is given
to the individual ship and not to the ship's owner. The permission
covers extraction in all Danish waters apart from a 300 m zone along
the coast and some other local limitations.
    The Raw Materials Act also allows for granting a concession to a
firm or consortium for a certain area of the sea. The concession can
contain requirements as to preliminary investigations, techniques of
extraction, use of the materials, treating the area afterwards etc.
Over and above this, the Raw Materials Act contains the usual
administrative rules for prohibiting or limiting extraction in
designated areas.

D. A. Ardus and M. A. Champ (eds.), Ocean Resources, Vol. I, 139–152.
© 1990 Kluwer Academic Publishers. Printed in the Netherlands.

Figure 1. Mette Miljø at seismic survey in Danish waters.

    The Conservation of Nature Act makes it possible to protect areas
of the sea. The conservation order can include limitations or a total
ban on e.g. the extraction of raw materials. Furthermore all ancient
monuments (wrecks, stone-age settlements etc.) that are over 100 years
old are protected.
    Additional large areas of Danish domestic waters have been
designated international protection areas under the Ramsar Convention
and the EEC Directive on Conservation of Wild Birds, Fig. 2.

Figure 2. Ramsar and bird protection areas in Denmark.

2.   EXTRACTION OF RAW MATERIALS IN SHALLOW WATERS

For the most part raw materials are extracted from Danish domestic
waters and only to a limited extent from the North Sea and the Baltic.
As can be seen from figure 3, Danish domestic waters are mostly shallow
with depths of less than 20 m. EEZ covers all Danish waters.
      For technical reasons raw materials are only extracted from depths
of less than 20 m and the extraction is often concentrated in depths of
less than 8-10 m.
      The areas of greatest natural and cultural interest such as
breeding and feeding areas for birds, reefs, wrecks of ships, stone-age
settlements etc. are to be found precisely at depths of under 8-10 m.
      This is why it is exactly in this area that conflicts arise but it
is not certain that the solution is to be found here.

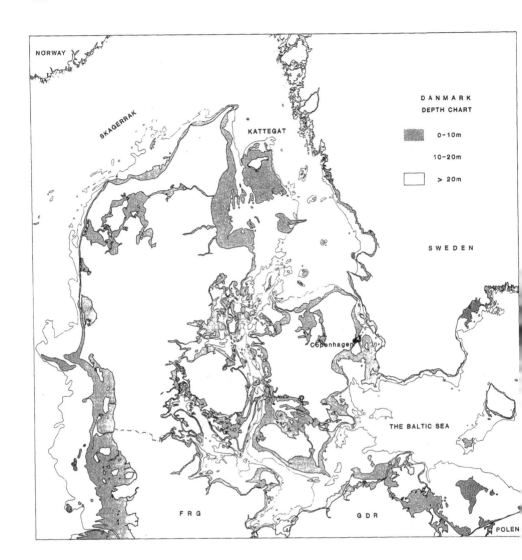

Figure 3.   Depth chart of Danish waters.

3.   RECONNAISSANCE ASSESSMENT OF RESOURCES AND CONDITION

The initial reconnaissance assessment of resources and conditions is
carried out by the Marine Raw Materials Survey in areas of 700-1500 km²
depending on geological and hydrographical conditions.   Investigations
are usually carried out in 3-5 areas each year.   The initial
reconnaissance assessment is in three phases:

3.1. Phase 1. Mapping and preliminary evaluation

A general geological survey of the area in question is carried out,
followed by a detailed survey of areas containing raw materials
(Fig. 4).

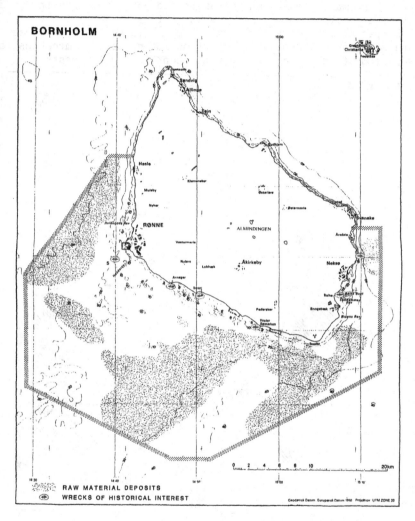

Figure 4. Raw material deposits offshore Bornholm.

The tests include profiling with shallow seismic equipment
(boomer, sub-bottom profiler, echo sounder, sidescan sonar), shallow
coring (vibrocore, gravity core, grab, jet-air) and laboratory tests of
selected samples.

The results, which are included in internal reports for each area,
contain a preliminary evaluation of the areas with raw materials.

3.2. Phase 2.   Balancing the interests.   (Evaluation of natural
     resources versus environmental and conservation interests).

An evaluation is made of the biological, environmental archaeological
and oceanographic interests in the area under investigation.   The
evaluations are mainly based on comparing and collating existing data
and only to a small extent on new investigations.
     By comparing maps of areas containing raw materials from phase 1
with maps of areas containing conservation interests, a comprehensive
view is gained of the areas that contain conflicting interests.

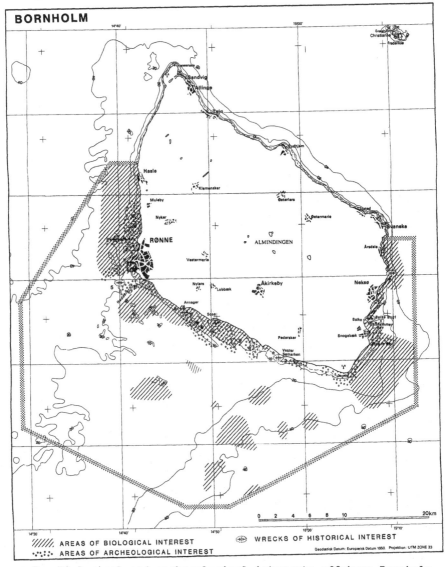

Figure 5.   Biological and archaeological interests offshore Bornholm.

   Then an evaluation of the interests is made. The result is a
survey of the areas where the raw materials are more important than
conservation interests, and areas containing raw materials where
extraction is to be limited or prohibited.
   The results of phase 2 are presented in public reports, "the Blue
Sea Reports".

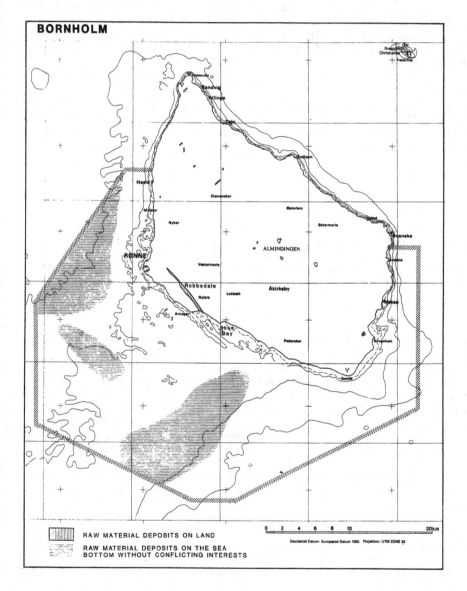

Figure 6. Raw material deposits with no or limited environmental
conflicts.

3.3. Phase 3. Final evaluation and exploitation tests.

A final investigation and evaluation is made of the areas containing
raw materials where there are no conflicting interests or where the raw
materials are more important than the conservation interests.
    This phase can include supplementary shallow seismic profiling and
shallow coring, but it is most frequently limited to taking samples
with a sand-pump dredger in order to evaluate the extraction conditions
and to obtain large samples for laboratory experiments (Fig. 6).
    The laboratory experiments include grain size analysis, organic
content and cement technological analyses (Potential Alkali Reactivity
of Cement etc.).
    The results are presented in a report which is available to the
public.
    As can be seen from Fig. 7, the geological survey covers about 70%
of Danish domestic waters. Phase 2 has been carried out and the
results published for about 25% of the area; the final phase, phase 3,
covers about 10% of Danish domestic waters. The final reconnaissance
and evaluation of all the Danish domestic waters is planned to be
completed in 1997.

4.   THE RESULTS OF THE MAPPING AS A TOOL OF ADMINISTRATION

The results of the mapping have been used to solve concrete conflicts.
In the following, three typical conflicts are described where the
traditional solution would have been a limitation of or prohibition on
extraction pursuant to the Raw Materials Act.

4.1. Case 1. Bornholm

About 90% of industrial minerals are excavated on land and only 10% on
the seabed. Excavation on land comes into conflict with other
interests such as nature protection, urban development and the supply
of water. The big sand and gravel excavation project in Robbedale on
the island of Bornholm can in the long run result in a shortage of
drinking water as the groundwater level becomes exposed because of the
deep gravel pits. Due to this the Marine Raw Materials Survey started
on a geological map of the seabed around Bornholm in order to find
industrial minerals which could replace those of high quality used in
the Robbedale foundries and glass works.
    The mapping has resulted in an agreement to reduce excavation over
a period of 4 years after which production will only take place on the
seabed.
    A small-scale extraction of raw materials for the cement industry
has been carried out for many years in Sose Bugt (Bay) close to the
coast. In Sose Bugt there are archaeological interests (wrecks and
stone-age settlements) and a danger of coastal erosion. This
extraction will be moved to the new areas west and south of Bornholm,
(Fig. 6).

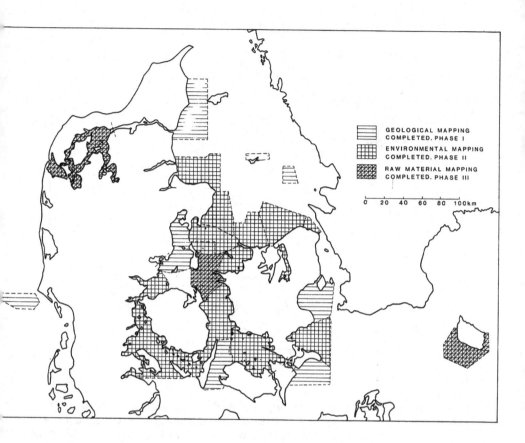

Figure 7.   Phase 1, Phase 2 and Phase 3 investigations in Danish
waters.

4.2. Case 2.   Anholt - Faxe Bugt (Faxe Bay)

In 1984 when the limit of territorial waters between Denmark and Sweden
was decided, the eastern part of the Store Middelgrund, an area rich in
industrial minerals, became Swedish.   This meant that Danish firms were
not allowed to dredge in the area to the same extent as previously.
This resulted in an intensifying of dredging on Anhold Østerrev, (the
eastern reef of the island of Anholt), as it lies in close proximity to
the former dredging area.   Unfortunately the reef has a large colony of
seals and many wrecks, including some warships and an old East-
Indiaman.   Anholt Østerrev has been very dangerous for shipping and it
is mentioned with awe in the old log-books.   The Marine Raw Materials
Survey has now mapped the waters around the reef, and in cooperation
with the dredging firms has designated a small area in which there are
no wrecks as an area where dredging may take place until an alternative
area may be found, figure 8, Marine Raw Materials Survey.

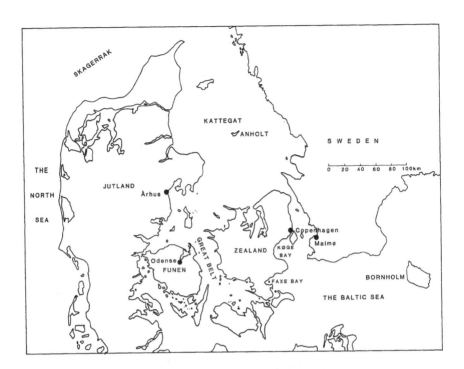

Figure 8.   Location map.

However, this is only a temporary solution. As the extraction that
takes place around Anholt is mainly used to supply the Copenhagen area,
it was decided to find alternative sources of raw materials closer to
Copenhagen.

On the basis of an initial reconnaissance assessment in the waters
around the Copenhagen area, in 1985 it was decided to carry out a
reconnaissance assessment of resources and conditions in Fakse Bugt.
The result was the discovery of sufficient raw materials of the same
quality as at Anholt and within shorter sailing distance of Copenhagen.
Already in 1988 very much of the extraction that used to take place at
Anholt Østerrev has been moved to Fakse Bugt without using the
regulating powers contained in the Raw Materials Act (Fig. 8).

4.3. Case 3. Køge Bugt (Køge Bay)

A Swedish car-producing company on Danish "soil". This sounds rather
strange! Apart from the fact that the "soil" now lies in Malmo,
Sweden, the project is already a reality. In 1987 the Swedish car-
producing company, Saab-Scania, lacked a total of 3 million m2 of sand
in order to fill in a harbour dock. The intention was that a test
track and a parking lot should be built in place of the dock in
connection with that part of the Kockum wharf which Saab-Scania had
taken over in order to set up a new car factory.

Based on the maps of the industrial minerals in Danish waters made by the Marine Raw Materials Survey, two areas could be found in Køge Bugt where it would be possible to dredge up large amounts of low quality sand without affecting the daily supply of raw materials to the capital. The National Forest and Nature Agency's mapping of the ecological and cultural/historic interests, supplemented by information from local fishermen, showed that the area was not a spawning ground of any importance for fish and that there were no shipwrecks in the area where the exploitation would take place (Fig. 8).

In January 1988 Danish firms terminated the delivery of the biggest export order of industrial minerals up to the present in record time (4 months ahead of schedule). This is one of the first examples of the Marine Raw Materials Survey's new policy; considerations about interests on the seabed are weighed quickly before permission is given for the extraction. This could not have happened without the systematic mapping of raw materials.

Danish experiences with the use of surveying in connection with the state's administration of the sea can be summarized as follows:

-      that mapping is important in order to carry out countrywide planning of supplies and to be able to evaluate the interests in any conflicts.

-      that the mapping of raw material interests must be balanced by the mapping of conservation interests in order to be able to evaluate the interests in any conflicts.

-      that it is important that the state authority itself has the equipment and manpower to be able to carry out and evaluate the mapping.

5.   EXTRACTION TECHNOLOGY

In Denmark the ships which today have permission for extraction in Danish waters use the technique of stationary suction hopper dredging.

In some extractions temporary permission for trailing suction hopper dredging has been granted to foreign ships.

The Danish fleet of stone and sand dredgers has not kept up with technological developments. Many of the ships are over 40 years old which makes extraction at sea uneconomical. Often they cannot compete with the quarries.

6.   MONITORING

In Denmark monitoring in earlier or present extraction areas has only been carried out to a limited extent.

Therefore there is only limited knowledge of the long-term effects of extraction and the factors that are of importance in the establishment of an extraction area.

The great extraction in Køge Bugt, (Case 3), will be the first area where the effects of trailing suction hopper dredging and stationary suction hopper dredging will be monitored in the coming years.

7.   TECHNOLOGICAL DEVELOPMENTS IN THE FUTURE

On the basis of the experiences of the last 10 years in Denmark, some areas where technological development is called for will be suggested.

7.1. Reconnaissance Assessment of Resources

The development of technical equipment and methods of measurement vary greatly. Shallow seismic equipment has almost not developed during the last 10-15 years; on the other hand radio positioning equipment has kept up with technological development.
    In the coming years development will take place in two areas:

1.   The reflected signal which is received from the seabed is only used to localize the geological interfaces. The "geological noise" which is reflected from the layer itself contains information about the composition of the layer which should be exploited.
      A reflectibility metre which could evaluate the geological noise of each layer on the bases of e.g. the amplitude, the phase shift or the frequency would be a valuable supplement.
      The advantage is that it would increase the exploitation of a signal which is in any case being received without using new energy sources.

2.   The development in the processing of digital data from deep seismic in oil/gas searches should be able to be used in shallow seismic.

      The broader frequency spectrum and the high sampling rate that follows from that, as well as the problem with reflection from the seabed in shallow waters, should be able to be solved by adjustments of software in the processing.

7.2. Technical Development of Sample-taking

The development of shallow coring equipment has been limited during the last 10-15 years. Some small adjustments have been made in already existing types but the basic principles are the same.
    When evaluating resources of raw materials there is a need to develop equipment that can take a sample of up to 1/4 m3 in sand and gravel layers quickly. It should be possible to use the equipment from a small boat without having to anchor so that 15-20 sample positions can be covered in one working day.

7.3. Reconnaissance Assessment of Biology and Archaeology

A traditional mapping of the biological and archaeological interests in Danish waters would take 25-30 years.

Therefore an initial reconnaissance assessment in Denmark is being reinforced with the help of knowledge of e.g. depths.

The archaeological interests consist of wrecks and stone-age settlements and the methods of investigation are shallow seismic and diving. The wrecks are concentrated at reefs and shallow areas along the coast or along the great sailing routes. Investigating these areas would be more rewarding than general investigations of open areas of the sea.

Stone-age settlements are always situated by the coasts of that time; for this reason it is important to know the coastline of the stone age. The stone-age settlements that now lie on the seabed because of a rise in the level of the sea, are important as they have not been the objects of human activities.

The biological investigations have focussed on describing the flora and fauna that are characteristic of different depths. This knowledge can be used to monitor the state of nature and the effect of human intervention.

Analysing the depth is one parameter for achieving a quick view but are there any other parameters?

7.4. Follow-up Treatment of Extraction Areas

It is important that the technology used in extracting be further developed.

In Denmark, areas where extraction has taken place have usually resembled a lunar landscape with an uneven surface full of large and small holes.

The extraction industry in Denmark is blamed for two things:

- Extraction near the coast arouses fears of coastal erosion.

- The lunar landscape left behind in the extraction area makes fishing difficult or impossible. At the same time it produces bad conditions for re-establishing the flora and fauna of the area.

The first problem is being solved as the mapping project finds alternative areas of raw materials further away from the coast.

The second problem can be solved by using the results of the survey and developing new extraction techniques that leave the seabed in a condition that satisfies everybody.

One solution could be to leave the extraction area as a declivity with evenly sloping edges and an even bottom some metres deeper than before the extraction took place.

8.   CONCLUSION

In Denmark about 90% of industrial minerals are excavated on land and
only 10% from the seabed. Excavation on land comes more and more into
conflict with other interests such as nature protection, urban
development and the supply of water. Resources on land are
disappearing. The counties, which are responsible for the mapping on
land, have indicated on different occasions that in many areas raw
materials are now in short supply.
     Therefore it can be expected that a rising proportion of the
excavation will take place on the seabed. For this reason it is
important that the Marine Raw Materials Survey's reconnaissance
assessment of resources and conditions continues and is improved.
     Finally, it should be pointed out that it is not just the
extraction of raw materials that can come into conflict with
conservation interests. The other uses to which the sea is put, i.e.
fishing, transport, leisure time activities, dumping of waste, must
also be analysed and evaluated vis a vis the conservation interests.
     On the basis of a survey of interests in the seabed and an
analysis of users and their effect on the seabed, it should be possible
to make sensible plans for the future activities in the area of the
sea.

THE BALTIC MARINE ENVIRONMENT AS A SOURCE OF AGGREGATES,
AND AS A RECIPIENT OF DREDGED MATERIAL

Boris G.L. WINTERHALTER
Geological Survey of Finland
Espoo
Finland

1.   THE BALTIC SEA

The Baltic Sea is the largest brackish water basin in the world. It
occupies a crustal depression in the Precambrian crystalline basement
of the Baltic Shield and is bordered in the south and south-east by the
Palaeozoic sediments of the East European Platform. During the
Pleistocene the whole area was covered by a continental ice sheet of
substantial dimensions. Due to crustal rebound following deglaciation
together with sea level changes in the world ocean the Baltic Sea has
evolved through various marine and lacustrine phases into the semi-
closed basin of today.
        The limited water exchange with the world ocean together with the
high degree of industrialization of the riparian states of the Baltic
Sea was turning the area into a "cess pool". The scientific community
reacted violently in the early 1970's finally leading to the signing of
a convention, known as the Helsinki Convention, in 1974. In 1980, all
Baltic Sea States (Denmark, Finland, German Democratic Republic,
Federal Republic of Germany, Poland, Sweden and USSR) ratified the
Convention. The Baltic Marine Environment Protection Commission -
Helsinki Commission (HELCOM), with its headquarters (secretariat) in
Helsinki was established to coordinate the work.
        The Helsinki Commission is an Intergovernmental Body preparing
recommendations of scientific, technical, administrative and
legislative nature to be unanimously approved by all the riparian
states. In accordance with the Convention the Contracting Parties,
i.e. all countries bordering the Baltic Sea, agree to take all
appropriate measures in order to prevent pollution of the marine
environment of the Baltic Sea Area. Great progress has been made in
decreasing discharges of harmful substances from land based source
areas as well as the detrimental effect of poorly controlled shipping
practices, including discharge of e.g. oily waters. Agreement is
currently being reached in restricting discharge of e.g. waste products
associated with offshore exploration and production of hydrocarbons.
Similar recommendations, ultimately regulating national activities will
evidently also cover other aspects of marine exploration and
exploitation. Sea dumping in general is prohibited by the Convention.

D. A. Ardus and M. A. Champ (eds.), Ocean Resources, Vol. I, 153–158.
© 1990 Kluwer Academic Publishers. Printed in the Netherlands.

It does, however, exclude emergency cases and dredged spoils dumped by permission from national authorities, provided they do not contain significant quantities of harmful substances specified by the Convention.

In addition to the HELCOM recommendations, most coastal states have environmental protection codes taylored to their local needs. Typical activities not yet covered specifically by HELCOM are those related to the exploitation of submarine sand and gravel deposits. The need for regulatory measures even though on a purely national level has become obvious. In view of this a marine sand project was started in Finland in 1983 and a similar project on the environmental impact of dumping dredged material is currently being established.

2.   SOURCES OF AGGREGATE MATERIAL

The bulk of the aggregate material exploited for various commercial purposes in the Baltic Sea States is genetically associated with the Pleistocene continental glaciation that covered most of northern Europe, ending some 10,000 years ago. The material, consisting predominantly of fragments detached from the Precambrian crystalline basement, was prior to deposition forcefully reworked mainly be glaciofluvial processes, i.e. transported and deposited by glacial meltwaters mainly as eskers, kames and marginal formations. The ultimate end product was aggregate material of improved hardness and durability. The glaciofluvial deposits on land, generally of limited exploitable size and number, were rapidly exhausted in the proximity to many urban areas.

Today, most large urban areas suffer from a severe shortage of exploitable deposits of aggregate material. The transportation distances are continuously increasing and inevitably raising the consumer price of a fundamentally inexpensive commodity to exorbitant and often quite unacceptable levels. The fact that the entire Baltic Shield area was, without discrimination between land and water areas, glaciated during the Pleistocene, is proof that commercially exploitable deposits of aggregates similar to those now depleted on land should be found beneath the present sea surface

The first to exploit this fact en masse were the Danes. The Swedes were soon to follow and the Finns are likewise becoming ever so much aware of the possibilities that lie in the submarine deposits. In fact, the small port city of Kotka, in the eastern part of the Gulf of Finland, a major user of marine aggregates extracted last year about 1 m cubic metres of sand and gravel and at least a second 1 m cubic metre of material will be dredged this year. Likewise, the city of Helsinki has applied for a licence to produce this year 2 m cubic metres of sand and gravel from a submarine glaciofluvial deposit located last year in the course of marine geological mapping conducted by the Geological Survey of Finland.

According to available information, so far only very limited amounts of submarine aggregates have been used in countries bordering to the southern and south-eastern coastal stretch, although considerable displacement of submarine soils resulted from the offshore

amber industry, which used to be a major offshore and coastal industry.
The depletion of the amber reserves seems to have slowed down this
activity during recent years.

It is obvious that a considerable increase in the exploitation of
submarine deposits of sand and gravel in the Baltic Sea is inevitable
in the very near future. With this in mind, we started in Finland,
some years ago, a project to evaluate the environmental impact of large
scale production of submarine sand and gravel.

3.   THE MARINE SAND PROJECT

In 1983 the Geological Survey of Finland together with various local
and government agencies started a research program designed to answer
at least some of the problems visualized to be associated with bulk
removal of large submarine soil masses. There had been cases where
dredging had been conducted too close to shore with beach erosion as a
result. Fishermen insisted that exploitation of sandbanks would be
detrimental to fish spawning. Fines dispersed in the sea water during
dredging was considered an environmental hazard.

The Geological Survey of Finland started the project by conducting
a detailed marine geological survey covering an area of about 1000 km²
off the town of Kotka in the eastern Gulf of Finland. Several
attractive glaciofluvial sand deposits were identified. The submarine
extension of an esker forming the island of Pitkäviira was chosen as
the area for a controlled dredging operation involving the removal of
100,000 cu.m. of sand and gravel by suction dredging during the summer
of 1985. A thorough study of the environmental conditions prevailing
within the proximity of the proposed dredging site was conducted during
the previous year to establish the conditions prior to intervention,
including such aspects as fish catches, fish population, spawning
habits, sea bottom ecology, water quality, current pattern, etc. The
environment was continuously monitored during the dredging operations,
and the next year a follow-up study of the conditions was performed.

The pilot dredging site was chosen with purpose to exaggerate any
adverse effects that the operation might have on the environment. An
area of 14 hectares not closer than 300 meters from the nearest shore
with depths ranging from 3.5 to 22.5 meters was staked out with buoys.
After the removal of the planned 100,000 cu.m. of sand, a 17.5 meter
deep, steep walled "gash" was formed in the side of the formation
covering an area of 3.5 hectares. The nearby sea bottom is deeper than
the deepest part of the pit, thus there will evidently be very little
sediment accumulation on the newly formed bottom.

During the actual dredging operation a very marked increase of
suspended fines in the close proximity to the dredger was observed.
All traces of increased turbidity downwind from the ship had, however,
vanished within a distance of no more than a kilometre or two downwind
from the dredger. No major changes in water quality parameters could
be observed during the dredging. All benthos at the dredging site and
its close proximity was, however, destroyed. The conditions at the
site were severely disturbed even a year after the operation with a
substantially lower biomass although the number of species had been

restored. No influence on fish catches nor on the spawning success in
the vicinity of the study site could be observed.

The final outcome of this first of several planned controlled sand
dredging operations was that the adverse influence on the environment
is negligeable provided that the site is chosen with care. Such
factors as the total amount to be recovered, the time and duration of
the operation, the quality of the material to be dredged especially in
relation to the content of fines, and other local conditions may
influence the final impact on the environment.

4.   MANAGEMENT AND LICENCING OF SEABED AGGREGATE DEPOSITS

Due to the new general awareness in ecological and environmental
matters, the actual code of practice to be applied in the future
management and licencing may change considerably as a result of
measures approved by the European Community and the Helsinki
Commission. At the present moment, however, a contractor applying for
a licence to extract mineral aggregates from the seabed needs
essentially only a permit from the owner of the sea area (viz. fishing
rights) and a consent from the State Water Court. The latter is given
only in the case that the proposed dredging operation is not in
conflict with other interests, nor does it adversely affect the
environment. If found necessary, environmental studies may be required
in the proposed dredging area and the ensuing costs can be charged to
the applicant.

The ownership of the seabed within the internal territorial
waters, i.e. the area between the outermost islands and the mainland,
is presently divided mainly between various private bodies, and to a
lesser extent, counties and the State. The outer territorial waters
(four nautical miles) are under the jurisdiction of the Ministry of
Agriculture and Forestry and the Exclusive Economic Zone is controlled
by the Council of State.

5.   MARINE DUMPING OF DREDGED SPOIL

Contrary to the recovery of mineral aggregates from a marine deposit
with low amounts of fine grained particles and dissolved substances,
the marine dumping of fine grained dredged masses from harbors and
other severely polluted areas must obviously have a greater influence
on the marine environment. Although e.g. the city of Helsinki has
monitored the quality of the surrounding sea waters with some of their
permanent stations located adjacent to a dredged spoil dumping site
that has been in use for several decades, none of the local city
officials, not to speak of the officials from the Ministry of the
Environment, seem to have a clear view of the environmental impact of
this dumping activity in an area of major recreational use, including
fishing.

With this in mind a new project aiming to increase our knowledge
on the environmental effects of marine dumping of dredged spoils has
been started. Approximately 1000 cu.m. of average dredged spoil will

be dumped in a properly documented spot. Sea bottom conditions as well
as water quality measurements will be conducted before, during and
after the dumping. It is hoped that the collected data will aid in the
formulation of legislative measures to be associated with future marine
dumping.

6.   THE MARINE GEOLOGICAL MAPPING PROGRAM

A prerequisite to all dredging operations is a thorough knowledge of
the local conditions with special emphasis on the geology and
topography of the sea bottom. The need for detailed marine geological
information is obvious. This challenge has been taken up by the Marine
Geological Group at the Geological Survey of Finland.
     The strategies involved in a detailed marine geological survey
program were first drafted in the mid-seventies. Due to the
irregularity and complexity of the Finnish coastal area with its
multitude of island and skerries it was decided to conduct the mapping
in the scale of 1:20000 utilizing the national grid system and the map
sheet division established for the adjoining land areas. An average
survey line spacing of 500 metres was considered appropriate for the
majority of areas although both wider and narrower spacings have been
utilized. Reliable and accurate positioning was deemed necessary. A
Motorola Miniranger system with data processing was chosen to do the
job.
     The "task force" at the Geological Survey of Finland operates
during the three to five summer months, a 13 metre survey launch -
r/v KAITA exclusively built for shallow water seismic and acoustic
profiling, and a 40 metre, 225 brt, converted road ferry - r/v GEOLA,
which is used for vibrohammer coring and as a floating base for semi-
realtime processing of marine data.
     The survey data, including echo-sounding, continuous seismic
profiling, and sidescan sonar imagery, collected by KAITA, is taken on
board GEOLA for immediate interpretation and evaluation. The
computerized graphic work station on board including two Hewlett-
Packard microcomputers (HP 9000/236C and HP 9000/310), a size A1
digitizing table and multipen plotter is utilized extensively in this
process. The preliminary information is used to choose suitable
sampling sites for vibrohammer coring using core barrels either 6 or 12
metres in length and normally with an outer diameter of 127
millimetres.
     The final computer drafted maps in a scale of 1:20000, each
covering an area of 10 km by 10 km, are in data bases and actual map
printouts are produced on demand to qualified end users. A total of
over 60 such map sheets are currently in the process of being stored in
the marine geological data bank. The first of a set of publicly
available marine geological maps is presently being prepared for
printing on a scale of 1:50000.
     Although the procedures adopted in the actual seabed mapping of
geological features are rather well established, the acoustic methods
used could surely be further improved to give a better combination of
penetration and resolution. Representative samples from sand and

gravel deposits cannot be satisfactorily acquired with available coring
equipment. Thus, actual production plans must often be based on
inadequate data or on test dredging implying high costs. A widely
acceptable code of practice for defining the permissible environmental
impact of both dredging and dumping should be established as soon as
possible to ensure proper use of our dwindling resources.

Figure 1. Sand dredger at work off the island of Pitkäviira, in the
eastern part of the Gulf of Finland. Note the plume of suspended fines
on both sides of the vessel moving with the wind-induced current.

A CASE STUDY OF THE MINERALS ON THE SEABED OF THE WEST COAST OF INDIA

S. RAJENDRAN
Cochin University of Science & Technology
Cochin
Kerala
INDIA

ABSTRACT. The occurrence and distribution of the mineral resources on
the western continental margin of India, excluding hydrocarbons, are
discussed by compiling all the available data. The margin is divided
into 4 zones viz: the north-west coast (Gulf of Kutch-Bombay), Konkan
coast (Bombay-Mangalore), south-west coast (Mangalore-Cape Comorin) and
the Laccadive area. The exploration and assessment of the mineral
resources are supported by geological and geophysical evidence. As
compared to the south-west coast, the heavy mineral deposits of the
Konkan coast contain more rock forming and unstable minerals like
augite, hornblende, epidote, zoisite etc. in addition to placer
minerals like ilmenite, rutile etc. The south-west and Konkan coasts
are promising for heavy mineral exploitation; the Laccadives for
calcareous sand and the Vembanad estuary and the Gulf of Kutch for
shell deposits. Offshore placer deposits of the Konkan coast are
fairly high and widespread compared to its onshore deposits, whereas
the offshore placer deposit potential of the south-west coast is yet to
be studied. The offshore region of Bombay-Quilon is suggested for
detailed exploration for phosphorite deposits. The terrigenous
deposits of Kerala and Konkan coasts can be mined by wire-line, bucket
ladder and hydraulic dredges depending upon the location and similarly
the biogenous deposits of Laccadives, Vembanad estuary and the Gulf of
Kutch by hydraulic suction dredges.

1.   INTRODUCTION

With the continuing exploitation of the reserves of the natural
resources from the earth's land areas for the rapidly increasing
industry, the resources which are finite, have been diminished
considerably. The need to win the same resources from the seabed, is
increasing day by day. India with a 5700 km coast line and proven
hydrocarbon resources within the Exclusive Economic Zone has a
tremendous stake in offshore exploration.

D. A. Ardus and M. A. Champ (eds.), Ocean Resources, Vol. I, 159–167.
© 1990 Kluwer Academic Publishers. Printed in the Netherlands.

2.   PHYSIOGRAPHY

The west coast of India is a typical example of a passive continental
margin.   In the east, the margin is flanked by a narrow zone of onshore
Tertiary sedimentary basin, formed as a consequence of down-faulting
and graben formation.   The regional fracture pattern, lineaments and
the rifting along the west coast produced the offshore sedimentary
basins and associated volcanicity in the Laccadives area (Sinha Roy and
Thomas Mathai, 1979).   Contrary to the east coast of India there are no
deltas here.   The western coast line is modified by head lands, bays,
lagoons etc. at irregular intervals.   The coast line shifted a number
of times.   Maximum sea level change of 150 m was recorded and several
submerged terraces can be marked at depths 110 m, 90 m, 85 m.   The
width of the continental shelf, varies from 300 km near Bombay to
approximately 80 km near Porbandar, narrowing further towards 60 km
near Cochin (Fig. 1).   The National Institute of Oceanography, Goa,

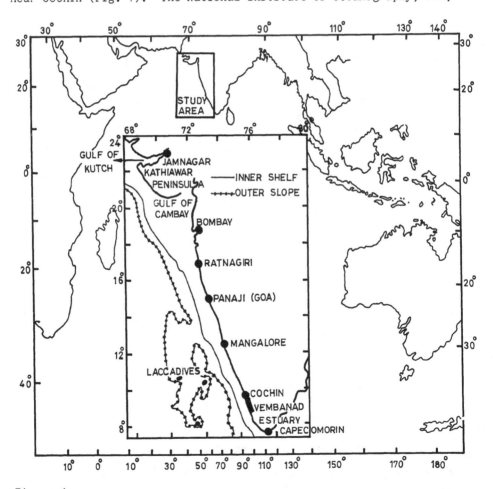

Figure 1

India (NIO) surveys show that the innershelf, to a depth of 60 m, is smooth and featureless. Seaward to a depth of 200 m the bottom becomes irregular by the presence of pinnacles (1-8 m height) and adjoining troughs that have originated through constructional and erosional processes operating upon algal and oolite carbonates of Holocene age (about 9,000 - 11,000 year, Nair, 1971). The shelf break is generally found at 130 m.

3.   GEOLOGY

The Indian peninsula is a shield area composed of Archaean and Precambrian igneous and metamorphic rocks. Faulting, broad uplifting and down warping have created sediment filled basins and shelves on the northern shelf area. The Kutch and Cambay basins, Kathiawar peninsula and the shelf off Bombay are regional depositional areas.

The continental shelf off Bombay has low relief with the shelf break occurring at about 90 m. The shelf is most probably underlain by Tertiary sediments. Seismic works suggest that Tertiary formations in the Cambay Basin extend south onto the shelf off Bombay (Sengupta, 1967). The surface rocks consist of lavas that are more acidic than the Deccan Trap basalts. The 1800 m thick Deccan Trap series abruptly terminate to the west, indicative of a down fault along the coast of Bombay. The western extension of this down faulted trap series lies beneath the shelf, according to Krishnan (1953) and Pepper and Everhart (1963). Along the southern part of the west coast Pleistocene and recent sediments are found in narrow belts. Traces of basaltic rocks, having similarity with Deccan Traps, have been found between the Laccadives and the Malabar coast. These are either an extension of the Deccan Traps which are subsequently down faulted by the west coast faulting, or a product of volcanic activity related to rifting and development of the west coast (Sinha Roy, 1979).

Over the entire width of the continental shelf, the sediment distribution pattern may be divided into two major types. The near shore facies up to 60 m consist of recent materials, being contributed by the present day rivers draining the west coast. The recent sediments consist mainly of silty clays and form a continuous near shore blanket from Bombay to Quilon. From a depth of 60 m it thins out and is replaced by quartz sand. The pattern of sediment distribution i.e. clays on the inner shelf and sands on the outer shelf, is a prominent feature of the western continental margin of India. The western flank of the Laccadive ridge constitutes a part of the Indus Cone sediments. The development of the west coast and the major phase of the sedimentation in both onshore and offshore areas are linked with regional tectonics of the Indian plate (Sinha Roy and Thomas Mathai, 1979).

4.   TERRIGENOUS DEPOSITS

Placer deposits of heavy minerals occur in discontinuous patches on the beaches along many parts of the west coast. Heavy mineral sands occur

(onshore) in area (Fig. 2) extending from Quilon to Cape Comorin in the
south-west coast. Calculations made by Siddiquie et al, 1984, reveal
that this part contains 17 million tonnes of ilmenite, 1 million tonnes
of rutile, 1.2 million tonnes of zircon and 0.2 million tonnes of
monazite. The richest offshore placer deposits occur along the 25 km
stretch of the south-west coast from Neendakara to Kayamkulam (9°10'N
to 8°58'N lat) in Kerala. The heavy minerals consist of kyanite,
sillimanite, zircon, garnet, ilmenite, leucoxene and rutile. The
highest concentrations of minerals are associated with the medium sand
which occur along the shore and in the southern part. The placer
minerals have been derived from the khondalites, granite-gneisses,
charnockites, pegmatite, nepheline syenites of the Western Ghats
(Mallik, 1985). They have been transported by the westerly flowing
rivers and brought to the present site of deposition by long shore
currents. The sorting action of currents and waves concentrate the
minerals. The present beach as well as part of the shelf was land
during Pleistocene as a result of the worldwide lowering of the sea
levels. There are a number of still strand phases that formed distinct
shorelines marked by beach ridges with heavy mineral stringers due to
regression and transgression in the Kerala coast. The monsoon activity
is vigorous in this part of India, the beach sediments as well as the
sediments brought by the river were reworked and the heavy mineral
pockets were concentrated in some areas only (Mallik, 1985).

Along the Konkan coast (19°N-12°N lat) the placers containing 14-
17% ilmenite form an estimated reserve of 4 million tonnes. The N10
surveys (magnetic, seismic, echo-sounding and seabed sampling) show
that the sediments off the Konkan coast consist largely of silty sand,
sandy silt and clay. The core samples show that a reddish brown clay
overlay olive grey clay which in turn overlay heavy mineral rich sands.
The heavy mineral concentration ranges up to 90% and has mainly
ilmenite and magnetite with minor quantities of apatite, rutile,
zircon, kyanite, etc. As compared to the Kerala coast, the Konkan
coast placer deposit contains many rock forming and unstable minerals
like augite, hornblende, epidote, zoisite etc. in addition to placer
minerals like ilmenite, rutile etc. The magnetite in the sands are
titanomagnetite or finely intergrown magnetite with ilmenite.
Siddiquie et al., reported 42-75% TiO_2 in the ilmenite. NIO's seismic
studies reveal that the first reflector is due to heavy mineral sands.
At water depth more than 10-15 m these are overlain by clays. The
heavy mineral bearing sands extend below the clays to water depth down
to 20 m. The heavy mineral sands vary in thickness from 2-10 m, but
thicker layers of sand (more than 10 m) are associated with river
mouths, buried ancient river channels and local sand lenses. The sands
are underlain by weathered basalt and in turn by the Deccan basalts
(Late Mesozoic - Early Tertiary).

NIO conducted magnetic surveys over this region, to correlate the
magnetic anomalies with heavy mineral concentrations, basement
configuration and the structural trends. Their work showed a number of
medium to high amplitude short wave length anomalies which were
interpreted as extension of heavy mineral placers below the seabed.
Concentration of ilmenite is higher in the offshore than the onshore.
Onshore deposits are restricted to a narrow coastal tract of beach sand

but the offshore deposits are fairly widespread. Their studies
concluded that the relict sands are terrigeneous. A systematic
investigation of shallow offshore region (south west coast) has been
initiated for getting a clear idea of the mineral resources by the
Centre for Earth Science Studies, Kerala (Mallik, 1985). The south
west coast's rich onshore placer deposits (ilmenite and monazite) make
the offshore sands promising for further explorations.

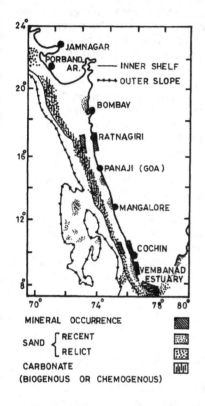

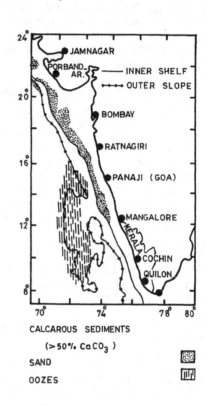

Figure 2   Figure 3

5.   BIOGENOUS DEPOSITS

The calcareous deposits are found in shallow areas of the continental
shelf and on atolls and banks (Fig. 3). The presence of shells within
the clay and sand horizons is an important feature in the Mangalore-
Cochin area (Sinha Roy and Thomas Mathai, 1979). The best known
deposits are in the Gulf of Kutch and in the Vembanad estuary (Fig. 2)
of Kerala. Biogenous sediments in the form of shells are being mined.
They have been formed by the accumulation of shells of living organisms
like Villoritta, a living species of lamellibranch. Earlier studies
estimated the shell horizon to be 0.5 to 2 m thick. Calcareous oolites
are found extensively in the outer part of the western continental

shelf of India. They often contain 7% of calcium phosphate. These
partly phosphatic oolites are found to contain appreciable quantity of
uranium and hence a possibility of being the source of the metal. The
calcium carbonate of these oolites is probably derived from upwelling
water which, at lesser depths, throw out part of the calcium carbonate
present in the solution. Sands with black calcareous oolites occur in
the bottom sediments in an elongated zone from the offshore areas
between Gulf of Kutch and Cochin ($23°-10°$ lat) running nearly parallel
to the present western coastline of India in water-depths ranging from
81 to 100 m and about 70 km off from the present coast. This zone
abruptly terminates at its eastern margin at a water depth of about
81 m (Mrinal K. Sen, 1987). The relict sands occur more than 50 m in
the shelf.

Deposits of coral and shell sands occur in the shallow areas and
lagoons of the Laccadives and appear to be promising economically.
Siddiquie and Mallik (1973) studied 11 of these lagoons and their
survey indicated 288 million tonnes of calcareous sand occur in the
lagoons in a 1 m layer. A further 423 million tonnes probably occur in
other lagoons where no surveys were undertaken. If the thickness of
the sands is greater as is to be expected, the reserves will be
considerably greater. The reserves between 1 to 2 m below the lagoon
floor are estimated to about 712 million tonnes. Recent surveys
conducted by the Geological Survey of India (GSI) show the seafloor in
the Laccadives is blanketed by clay and calcareous ooze; however in the
shelf portion and around Laccadive ridge sand/silt is also observed.
No hard rock was recovered. Calcareous sand dominates around the
Laccadive-Chagos ridge. Chemical analysis of seabed samples reveal
that zinc, nickel and phosphate contents are less in deeper water
sediments (GSI Newsletter, 1987).

6.   CHEMOGENOUS DEPOSITS

The occurrence of phosphate, barium concretions and manganese
encrustations have been recorded from the western continental shelf and
slope of India. The P_2O_5 content of phosphorite nodules of the west
coast vary 20-30%. The upwelling and nondepositional environment is
similar to regions like Peru-Chile, S.W. Africa (Rao et al, 1987).
Initial surveys of Rao et al, 1978 showed that the phosphate content of
the sediments is only 0.1-1.3%. NIO's geophysical surveys between Goa
and Bombay indicate NNW-SSE trending ridges rising to a height of 2-5 m
above the seabed. These ridges are algal, coral and shell limestones
and extend out from the clay sand boundary at a depth of about 50 m to
the shelf edge, and have been traced over a distance of over 150 km
along the shelf. Rock samples from these ridges show that some contain
up to 5% phosphate (Siddiquie et al, 1984). Algal and oolitic
limestones were recovered from depths ranging from 70-150 m from $13°N$
to $8°N$ lat (Nair, 1985). The work carried out by Rao et al, 1987
clearly shows that in the inner shelf, the phosphorous is mainly
associated with iron and organic carbon. In the outer shelf/slope its
association with carbonate suggests its diagenetic origin. These
studies suggest that the outer shelf/slope region between Bombay and

Quilon is promising for detailed exploration. Earlier work on
phosphorite along the western shelf revealed a specimen containing
about 10% P_2O_5 found in 1967 by a Russian Oceanographic Expedition at a
sampling station in the slope (200-300 m depth) off Quilon (Baturin,
1981). Murphy et al, 1968 and Marching, 1972 reported the distribution
of phosphate in the core samples collected from a few transects across
the continental margin between Alleppey, Bombay, Cochin and Karachi.
Nair (1969) reported phosphatised ooids from the outer shelf region
between Bombay and Mangalore. Setty and Rao (1972) studied the
distribution of phosphate in the cores collected from the shelf and
slope regions off Saurashtra coast. Rao et al, 1978 suggested that the
outer shelf region between the Gulf of Cambay and a little south of
Bombay and the continental slope region are environmentally favourable
for detailed operation of phosphate.

7.   TECHNOLOGY FOR EXPLOITATION

Terrigenous Deposits: The technology for exploitation of the
terrigenous placer deposits on the continental shelf is already
available and is within the capability of the Indian industry. The
legal frame work of the deposit is also well defined. Offshore placers
of the Konkan coast can be mined by using wire-line, bucket-ladder and
hydraulic dredges. Wire-line dredges may employ drag buckets or
clamshell and hydraulic dredges, or suction or airlift principles may
be employed. The dredges can be located on flat bottom or catamaran
barges. Dredging of very near shore placers may be achieved by
operating from land. The bucket-ladder type of dredge has greater
digging ability and hence can be deployed for mining placers which are
partly consolidated or cemented or where the most valuable part of the
deposit lies at the contact between the alluvium and the bedrock. This
type operates very economically up to 45 m depth and needs minimum
power requirement per unit dredged. Suction type hydraulic dredges can
be operated very effectively up to 60 m water depths. This system is
quite ideal for the inner shelf. The maximum practical economic
working water depths for wire-line dredges are about 150 m. With the
drag bucket modified into large net buckets, the method can be utilised
to greater depths, at least on a pilot scale. They can be used in
areas of high water currents and swells. The type of bucket can be
changed readily with change in the nature of alluvium of substratum.
The hydraulic suction dredge is the most suitable system for mining
unconsolidated sediments. When the sediments are semi-unconsolidated,
a cutter head can be mounted on the lower end of the suction pipe to
agitate the seabed.
      Biogenous and Chemogenous Deposits: since these deposits occur in
unconsolidated and freely flowing states they can be mined by hydraulic
suction dredge which seldom requires a cutting head. The calcareous
sands of the Laccadives area can best be mined with a hydraulic dredge
on a stationary barge and the dredged materials can be transported in
large hopper barges to the shore. Barium and phosphatic nodules and
bone beds can be exploited in the same fashion as that of manganese
nodules. A narrow diameter air-lift dredge is ideal for dredging

calcareous sand and oolites from the continental shelves. The
associated clay and other fine detritus would be automatically washed
away while mining.

In the Vembanad estuary, the shells are dredged for the
manufacture of white cement. A cutter section dredge is being worked
in some parts of the estuary up to 10 m. The lime shells in the
Vembanad estuary are gathered manually using small country boats. The
biogenous deposits in the Gulf of Kutch (calcareous sand and shells)
are being exploited for cement manufacture. The annual production is
about 1 million tonnes.

8.   ENVIRONMENTAL EFFECTS OF MARINE MINING

The increased offshore production of oil in the west coast causes an
inherent risk of contamination of marine environment which can be
minimised by proper monitoring of drilling and transport. Calcareous
sand mining in the Laccadives area will create serious beach erosion
problems which will have adverse effects on the islands itself. West
coast of India has proved as one of the major monazite producers of the
world. Mining of beach zones requires additional attention in view of
its effects on littoral processes and erosion. Some parts of Kerala
coast already have erosion problems. Offshore dredging will have
definite impacts on the erosional and depositional equilibrium in the
area causing disturbance on the seafloor, destruction of benthic
communities, disturbance at the surface and within the water column due
to dislodgement and transport of mined material.

9.   REFERENCES

Annual Report (1985) National Institute of Oceanography, Goa.

Baturin, G.N. (1971) 'Formation of phosphate sediments and water
    dynamics`, Oceanology, Vol. II, pp.373-376.

Class, H. et al (1974) 'Continental Margins of India`, The Geology
    of Continental Margins.

Cronan, D.S. (1980) 'Underwater Minerals`, Academic Press, London.

Griffin, J.J. et al (1968) 'The distribution of clay minerals in
    world ocean`, Deep Sea Research, Vol. 15, pp.433-459.

Harbison and Bassinger, 'Marine Geophysical Study off Western
    India`, JGR, Vol. 78, No. 2, pp.432-439.

Kidwai, R.M. et al, 'Heavy minerals in the sediments on the outer
    continental shelf between Vengurla and Mangalore, West Coast of
    India`, Jour. Geol. Soc. India, Vol. 22, pp.32-38.

Krishnan, M.S. (1953) 'Geology in India and Burma`.

Madhusudan Rao, C.H. (1978) ′Distribution of $CaCO_3$, Ca^{2+} and Mg^{2+} in sediments of the northern half of western continental shelf of India`, Indian Jour. of Mar. Science, Vol. 7, pp.151-154.

Mallik, T.K. (1985) ′Marine Geological Studies on the Coastal Zone of Kerala`, Coastal Zone, pp.1571-1586.

Mrinal, K. Sen. (1987) GSI Seminar Abstracts.

Murty et al (1980) Indian Jour. Mar. Science, Vol. 9, pp.56-61.

Popov, V.P. (1978) ′Oceanology`, Vol. 18, pp.177-180.

Rao et al ′Distribution of phosphorous and phosphatisation along the western continental margin of India`, Jour. Geological Soc. of India, Vol. 30, pp.423-438.

Sinha Roy, S. and Thomas Mathai (1979) Professional Paper No. 5, Centre for Earth Science Studies.

Siddiquie et al (1984) ′Superficial Mineral Resources of the Indian Ocean`, Deep-Sea Research, Vol. 31, pp.763-812.

PART IV

Ocean Renewable Energy

RENEWABLE ENERGY FROM THE OCEAN

A. GORDON SENIOR
Gordon Senior Associates
Normandy
Surrey
England

ABSTRACT. The ultimate sources of renewable energy from the ocean are
the sun and the moon. The wind, flowing over the surface of the sea,
produces waves in proportion to the wind speed, duration and the fetch.
The moon's gravitational pull moves immense masses of water causing the
tides; while solar radiation pours down onto the surface creating warm
surface layers and vertical temperature gradients. All are
interdependent and create a constantly moving mass of water equal in
volume to several times the total volume of the land mass above the
water line.

1.   THE NEED

Man has used energy from the ocean for thousands of years - since the
days of the first sailing ships propelled by the wind, the colonisation
of Polynesia by an ancient South American people using ocean currents
to transport their rafts thousands of miles (as demonstrated by Thor
Heyerdahl in the Kon Tiki expedition), and by the early Chinese using a
primitive form of tidal power. For many centuries all the energy used
for sea transport came from the wind until the development of steam
power from coal fired boilers. In the last 150 years energy supply has
been totally overshadowed by the development of land based power
systems notably steam power from coal, and then petroleum products
refined from crude oil were used to replace the horse for road
transport, by hydro electricity and more recently by nuclear fission to
produce base load power with the future promise of cheap, limitless
power from nuclear fusion.

   The demand for fossil fuels - coal, oil and natural gas - while
not yet outstripping supply, has caused some concern that these finite
sources of energy may one day cease to meet our needs. Twenty years
ago the Club of Rome drew attention to the reckless use of limited
resources which would create future problems, but little has been done
to heed it. In the last fifteen years we have lived through three
world energy crises based on the traded price of crude oil and yet once
again we are awash with a surplus which suppresses the price of crude,
a state of affairs which is likely to remain until well into the next
decade. That is tomorrow's problem you say and so it may be; but if we

D. A. Ardus and M. A. Champ (eds.), Ocean Resources, Vol. I, 171–181.
© 1990 Kluwer Academic Publishers. Printed in the Netherlands.

fail to plan now for some redistribution of energy supplies or make
dramatic changes in energy uses, including conservation, then we will
be plunged for a fourth time in a quarter of a century into the dilemma
of a sudden increase in price of this basic commodity vital for the
support of our way of life. And the next crisis may be more serious
than those we have experienced to date. In 1973 when crude oil was
trading at $1.50 per barrel a price of $10 per barrel was unthinkable,
but it happened within a single year and now this figure, if it were
sustained, would be so low as to threaten the entire international oil
industry and ruin the economies of the OPEC producers. Is it
unthinkable that crude will trade at $60, $80 or even $100 per barrel
sometime during the next decade? We really must prepare for the future
which will be with us sooner than we think.

The highest national consumer of energy is the USA with an average
of 12.8 long tons (14.3 short tons) of coal equivalent per person per
annum. With 5.3% of the world population the USA consumes 44% of the
world's gasoline and 33.9% of the electricity.

Mankind's demand for energy is insatiable as the quality of life
improves in the developed world and the underdeveloped world tries to
catch up. It never will of course but this will not stop it wishing
and striving to do so; but to do so would mean that the world's known
reserves of realisable fossil fuel would be consumed within a few
months. Of course this is a preposterous thought but it may help us to
put our future needs of energy into perspective. The demographic
prediction is that the world population will double from 5 to 10
billion people by the late 2020's and it is not unreasonable to guess
that world demand for energy will increase at least four fold over this
same period.

Meanwhile, reserves of fossil fuels are being depleted at an ever
increasing rate; the "Greenhouse effect" and the holes in the ozone
layer are causing growing concern for our future because of the
changing climate and the increase in ultra-violet radiation which these
effects produce. The President of the Royal Society, in his
Anniversary Address in London in November 1988, said

> "During the year, acid rain and the carbon dioxide greenhouse
> effect vied with Chernobyl and Sellafield as public enemy number
> one. But fossil fuels and nuclear energy are the only sources we
> have at present large enough to supply our energy needs. Most of
> that energy is not a luxury - without it we could not survive. We
> therefore have no alternative but to live with the risks and
> hazards of energy production until, by more research, they are
> reduced or alternative sources are found. Here, and in all
> problems of this kind, people must understand that life is a
> gamble, risks are inevitable and all we can do is to play for the
> best odds."

Hopefully, in the future, we can improve these odds. Renewable energy
is all around us and most be explored and, where suitable, exploited.
The ocean provides a vast energy resource capable, theoretically, of
meeting all our demands if only we can find effective ways of
extracting it. It is clean, produces no toxic by-products and should

be environmentally more acceptable than conventional sources. Herein
lies a challenge.

2.   SOURCES OF RENEWABLE ENERGY

The main sources of renewable energy from the ocean are listed below:

    wind
    currents
    waves
    tides
    thermal gradients - OTEC

to which I shall add with an explanation which will follow

    biomass
    geothermal - hot rocks
    core methane
    solid methane hydrate.

The technical papers presented on Ocean Renewable Energy Utilisation
deal with waves and OTEC. I shall not conduct a detailed review of all
energy sources listed above but touch on some of those issues which I
feel are of interest.

2.1 Wind Turbines

Wind Turbines provide intermittent power depending on natural factors.
They may be of the horizontal or vertical axis type. Units of 100 kW
to 300 kW are now in service on land but the most effective new
machines are considered to be in the range 500 kW to 1000 kW. A
3000 kW unit is under trial in Orkney, Scotland. Wind turbines are
large, noisy, visually obtrusive and require a large space. A wind
farm in California illustrates what can be done where 1500 MW is
generated from turbines of 85 kW average. They occupy 77 square miles
of desert. Development has not reached the end of the road and perhaps
we shall see 5 MW units in service within the forseeable future.
Proposals have been made to locate wind farms offshore in shallow water
where the environmental impact (noise and visual intrusion) may be
less, but this must be offset against greater foundation costs, greater
access costs and navigational hazards. In the UK the Central
Electricity Generating Board (CEGB) is to build an offshore turbine
which is believed will be the first anywhere in the world. It has been
estimated that a 2000 MW wind farm would require about 400 square miles
of space. If the full resource available around the coast of Britain
were to be exploited an estimated potential of 230 TWh per annum could
be achieved from offshore turbines, which is roughly equivalent to the
whole of the UK power demand; but wind turbines do not produce base
load power. The technology is well advanced with scope for continuing
development.

2.2 Ocean Currents

Ocean currents occur in much of the sea mass and constitute another
major potential source of energy. However, harnessing of energy from
the currents requires anchorages to resist the force of the current
which precludes locating these devices in deep water. The world's
strongest currents are the Nakwakto Rapids, Slingsby Channel, British
Columbia, Canada where the flow rates may reach 25 feet per second. In
the Pentland Forth between Orkney and the mainland of Scotland speeds
of 17 feet per second are recorded and many other sites adjacent to
land have speeds in excess of 12 feet per second. In the Straits of
Messina, which separate the island of Sicily from Italy, the strong
currents reverse with the tides.

2.3 Wave Energy

Wave energy was first proposed in 1799 by two Frenchmen, father and
son, who filed a patent for a wave energy system to drive ratchet
pumps. A float was located in the sea by a beam connected to a fulcrum
on a headland; the beam would oscillate with the waves producing
angular motions to drive the pumps. The oil crisis of 1973 served as a
stimulus for deeper study which has been undertaken in a number of
countries, notably the UK, France, Norway, Portugal, USA, Canada,
Japan, Australia and now India. The most comprehensive wave power
studies have been conducted in Britain between 1976 and 1983
culminating, between 1981 and 1983, when seven different device teams
were commissioned to develop and present costed proposals for devices
capable of generating 2000 MW off the West Coast of Scotland. These
attracted colloquial names after their inventors or descriptions of how
they worked, thus:

-   The oscillating water column designed by the National Engineering
    Laboratory.

-   The Lancaster flexible bag of Michael French from the University
    of Lancaster.

-   The Lanchester Clam designed at Lanchester Polytechnic and
    developed by Sea Energy Associates.

-   The Cockerill raft named after Sir Christopher Cockerill who
    invented the hovercraft.

-   The submerged Bristol cylinder designed by David Evans.

-   The Belfast turbine designed by Professor Alan Wells.

-   The Edinburgh Duck designed by Professor Stephen Salter.

Very serious study and evaluation was undertaken by dedicated teams of
engineers, mathematicians and scientists from the universities,

consultancies and industry, whose enthusiasm was maintained through the period of review.

A detailed evaluation of the results of these studies led to the conclusion that wave energy was a long term and speculative proposition and, as a result, the programme was terminated. With hindsight one can question the sensibility of chosing 2000 MW as a power target when nothing within several orders of magnitude of this had been built. One can also question the proposed site location off a remote part of Scotland where the combination of high energy in the waves with very low local electricity demand necessitated this electrical power having to be transmitted over long distances and much of it exported into England. The location and the scale presented a challenge to the device teams who responded valiantly.

Little further work has been funded in the UK from public sources since 1983.

Meanwhile, Norway has built two demonstration wave power stations, an oscillating water column device built by Kvaerner Brug and a tapered channel device built by the Norwave Company, both of which have been generating electricity since November 1985. Unfortunately, the Kvaerner Brug station was damaged during a violent storm over the New Year weekend and will have to be rebuilt.

Happily wave power is not dead in Britain as the Department of Energy has now supported the building of a prototype which combines the oscillating water column and the tapered channel features of the two Norwegian devices. It has been designed by Professor Trevor Whittaker from the University of Belfast and is being built in a natural gully on the shore line of the island of Islay off the West Coast of Scotland. It will use a new Bi-plane Wells Turbine and have an installed capacity of 180 kW. The total cost has been estimated at £140,000 and will produce electricity for as little as 4 to 5p per kWh.

I will return to the subject of wave energy later.

2.4   Tidal Power

Tidal power is generated by collecting rising tidal water behind a barrage and then releasing it on the ebb tide through turbines to generate electricity. The power available is intermittent and depends on the lunar cycle which, unlike wind power, can be predicted. Tidal power has been used for centuries and there are a number of examples on coastlines of early tidal mills which were used for milling corn before the advent of steam power and which were competitive with land based windmills. The effectiveness of tidal barrages increases with increasing tidal range and this limits the number of suitable locations. Tidal barrages are producing power at several sites around the world including the USSR, Canada and France. The French scheme has a barrage across the Rance in Brittany and has operated successfully since 1968 generating 240 MW. Tidal barrages have been under study in the UK since the 1930's notably across the Severn Estuary between England and Wales where the mean spring tidal range is 40.7 feet or approaching three times the average around the coast of Britain. the Severn Barrage Tidal Power Group anticipate that the installed capacity of their project will be 7200 MW with an annual output of 14.4 TWh

which is equal to 5% of the annual demand in England and Wales.
Another British Group is studying a smaller scheme across the Mersey
below Liverpool.

While tidal power is clean the barrage may concentrate pollutants
upstream and alter the ecological balance, and this is a cause of
concern the the environmentalists.

2.5 Thermal Gradients

Thermal gradients along a vertical profile provide another potential
source of energy, known as OTEC or ocean thermal energy conversion.
These plants work by extracting energy from the temperature difference
between the warm surface layers and the cold deep mass beneath by the
use of heat exchangers. They can be floating or land based, and
operate on open or closed cycles. With present day technology, to be
economical, they need a temperature difference of at least 20°C
separated by not more than 1000 m in depth and therefore are only
applicable to tropical and sub-tropical regions. They present a number
of technical challenges such as the design and construction of the cold
water pipe through which cold water is drawn from the deep sea to the
surface, and the design of heat exchangers efficient at these
relatively small temperature differences.

OTEC is environmentally clean and, unlike wind, wave and tidal
power, generates power which is available continuously for 24 hours of
the day and thus can be a reliable source of base load power.

There is considerable interest in OTEC throughout Europe despite
the total absence of suitable locations within territorial waters. The
first proposal for exploiting the temperature difference in the water
column was made in France over 100 years ago and the French started
working on OTEC in the 1930's and again in the 1960's. Hawaii is an
ideal site for OTEC and perhaps not surprisingly, the world's first
experimental plant is located here where it tests pumps,
instrumentation and the equipment's susceptibility to marine bio-
fouling and corrosion.

2.6 Marine Biomass

Marine biomass can be harvested, dried in the sun and burned in
conventional power plants. This has yet to be employed on a large
scale but could provide a useful source of power to small island
communities. Initially it will be labour intensive and may be
difficult to control.

2.7 Geothermal Energy

Geothermal energy, or energy in the heat of the Earth's core, is,
strictly speaking, not renewable but is available in such abundance
that its large scale use will involve no significant depletion of the
resource. The heat can be extracted either by drawing up naturally
occuring hot water from underground reservoirs called aquifers, or by
pumping cold water down bore holes and passing the hot water which
returns to the surface through heat exchangers to extract the heat.

Cavities or widened fissures deep down the bore holes accelerate the heat transfer from the hot rock to the water. Experiments are being conducted on land in the UK in Cornwall where a hot spot is known to exist. These hot spots, which are locations where the temperature is higher at a given depth below the surface, also occur offshore. The technology has yet to be proved but it is anticipated that useful heat will be extracted in the future.

2.8 Core Methane

Core methane is a source claimed by Professor Thomas Gold of Cornell University, Ithaca and has yet to be proved. Professor Gold predicted some years ago that prodigious quantities of methane would be found saturating the Earth's deep rocks, an idea which has been ridiculed by many geologists. Gold had hoped for confirmation of his theory from a deep bore hole drilled into the base of an ancient meteor crater in Sweden, in rock known to have no hydrocarbon deposits. Although traces of hydrocarbon were found, the quantity was insufficient to convince those who disagreed with Gold and the issue remains unresolved. If Professor Gold is right then a vast limitless reserve exists and many preferred sites for its extraction will be under water.

2.9 Solid Methane Hydrate

Solid methane hydrate is another finite resource but available in such large deposits under the sea and land that it is included here as an ocean resource. Estimates for the total reserves have been made which are as high as 500 billion billion m^3 of methane which is sufficient to meet the world's present energy demands for 5000 years.

Solid methane hydrate is a chemically bound mixture of methane gas and water which occurs in vast sheets or lenses first found in the frozen wastes of Canada and the Siberian Artic and now has been found in many other parts of the world including the Gulf of Mexico, the Middle America Trench off the Mexican coast and the Black Sea, and on the deep ocean floor off the coasts of South America, Africa, Australia and the Far East, in the Pacific, Atlantic and the Caribbean. Russian geologists have estimated that 85% of the deep ocean floor is suitable territory for the formation of hydrates. The combination of pressure and temperature found at the bottom of the sea is suitable for hydrate formation on continental slopes below 1600 feet. The petroleum industry has known about solid methane hydrate for many years and regards it as a hazard, not only when drilling for oil and gas but also during preliminary processing of the crude at the well head where hydrate formation has to be avoided.

Gas from hydrates has been produced from the Messoyakh field in the USSR since the 1960's by pumping methanol into the hydrate zone and the peak production rate has been reported to be 6.5 Bcf/year. Where the ground is not frozen the gas can be produced by lowering the pressure around the hydrate sediments which allows the gas and water to dis-associate without the application of heat.

3.   SPIN-OFF BENEFITS

May sources of energy have spin-off benefits. Thus conventional power
stations produce large quantities of low grade heat which can be used
in combined heat and power schemes. So it is with the renewables.
Thus, some wave energy devices can double as breakwaters. OTEC plants
bring rich nutrients to the surface from the deep cold which can be
used to promote fish farming. Tidal barrages can provide estuary
crossings. Navigation lights can be fitted to wind turbines on sand
banks as a warning to shipping.

4.   COSTING

Comparative costings of energy from different sources, finite or
renewable, present a minefield of difficulties in reaching agreement
between different interested parties. It is common to think in terms
of capital cost per kW installed, and cost per kWh generated, but even
these figures are notoriously difficult to find general agreement. One
reason for this is that comparative costs place different energy
sources in competition with one another and competition produces
disagreement. Strict impartiality and the judgement of Solomon are
needed, and these are desirably qualities rarely found.

In the long term it is not in anyone's interests to distort the
figures in favour of one scheme or against another. It is important to
assess the total life cycle costs which include research, development,
prototype testing, design construction, operation, maintenance,
decommissioning and dismantling. It is also important to assess the
energy payback period, or the number of years which a scheme has to
operate before the total energy used in its creation will be replaced.
There are some inefficient power plants which may never achieve this
and should therefore never have been built.

5.   WAVE ENERGY IN BRITAIN

I said that I would return to the subject of wave energy and I now wish
to refer to the British Wave Energy Programme which was terminated in
the summer of 1983. Between 1976 and 1983 the British Department of
Energy supported a major programme of study into wave energy,
culminating in the spending of 16.5 million pounds sterling ($30
million) on the design and evaluation of conceptual schemes by 7
independent teams of academics, consultants and industrial firms for a
2000 MW installation to be located off the West coast of Scotland. A
leading form of consulting engineers was appointed to co-ordinate the
work and prepare a series of reports and, in turn, their work was
managed by a smaller team of specialist engineers and scientists
working as the Energy Technology Support Unit, ETSU for short, from the
Atomic Energy Research Establishment at Harwell. The conclusion drawn
from these reports was that the development of wave energy was so long
term and speculative that the programme should be abandoned until more
promising information was available. This result was disappointing

enough for the British Wave Energy community but worse was to follow.
A direct outcome of this decision, and information passed in confidence
to Brussels, was that the European Commission withdrew wave energy from
its list of activities worthy of funding; and now I understand that the
US Department of Energy has cited the "perceived" failure of the
British wave energy programme as a reason for supporting only ocean
thermal energy conversion. All efforts to have the British programme
restarted, except on a very limited scale, have been to no avail.

However, evidence was submitted to a House of Lords Select
Committee in April last year that for at least one of the devices, the
Edinburgh Duck developed by Professor Stephen Salter, this opinion was
not universally held by those who had studied the device in great
detail and that the conclusions of the first report drafted had been
modified, without consultation, before submission to the Department of
Energy. You could say that the Department had been misled. Their
Lordships took note of this and in their report, published on 21st June
1988, referred to "A serious conflict of evidence which only an
independent review could resolve".

This report was the subject of a major debate in the chamber of
the House of Lords on 1st December 1988 when support for an independent
review was made repeatedly by a number of speakers who had served on
the Committee; none objected or abstained.

I will limit myself to three short quotations from the official
report of this debate published in Hansard from which you can draw your
own conclusions. Opening the debate, Lord Renwick said

"Far more worrying ... was the sudden termination of support for
the wave power programme led by Professor Stephen Salter of
Edinburgh University. Professor Salter and others presented to
the committee evidence which was in serious conflict with the
official view. Since the publication of our report the committee
has received further evidence to support Professor Salter's claims
that the Department of Energy and ETSU handled his projects in an
unsatisfactory manner. Some serious allegations have been made
.... **We therefore call for an independent review of the whole
problem to ensure that wave power has a fair chance of serious
consideration."**

The Earl of Lauderdale, speaking from the government benches, said

"We found more than a mere whiff of intellectual and professional
skullduggery about the Department of Energy's decision to end
offshore wave power research."

Lord Rea referred to

"a 'massaging' of technical information, so that wave energy
appeared considerably more expensive than the other technologies,
especially wind power".

In reply the Minister, Baroness Hooper, said

"I am satisfied that all those involved acted responsibly and
completely refute any suggestion of cooking the books.

Looking forward, I have agreed with Professor Salter that there is
little to be gained by reviewing events of five to eight years ago
and we should concentrate on the future".

We live in hope but nothing yet has been done to restore public support
for wave energy research in Britain other than the support of Trevor
Whittaker to which I have already referred. I make no apology for
referring at some length to what might seem to be a domestic matter
because it has had a far reaching effect on other national attitudes to
wave power, in particular the USA and the European Community. I urge
those in positions of influence over national spending to look again at
wave energy and form their own conclusions from all the evidence and
informed opinions available to them. And I urge those protagonists of
wave energy not to lose heart but to redouble their efforts to find
ways of capturing effectively, economically and safely some of the vast
replenishable store of energy from the waves.

6.   THE FUTURE

Predicting the future is problematical even when extrapolating from
known technology. But technology makes large leaps when new
discoveries are made and these are much more difficult, if at all
possible, to predict. And technology affects only the supply side of
the equation; to have the total picture we have also to predict the
demand side. The only thing we can say with certainty is that firm
predictions can be notoriously unreliable. For example it was
predicted in 1900 that the growth of horse drawn traffic in London
would be so great that by 1920 the streets would be covered to a depth
of 6 inches by horse dung unless more efficient methods could be found
for its removal. No such methods were found but the problem vanished
with the development of the internal combustion engine.
     The future will be full of surprises but let us respond to these
surprises as challenges and not be overwhelmed by them.
     Many of you will have read the children's books written in the
last century by the Cambridge mathematician Charles Lutwidge Dodgson
who wrote under the psuedonym of Lewis Carroll. In his book 'Alice
Through the Looking Glass' Alice, when challenged by the White Queen,
said "I can't believe that", to which the White Queen replied in a
pitying tone "Can't you? Try again; draw a long breath, and shut your
eyes". Alice laughed and said "There's no use trying, one can't
believe **impossible** things". The White Queen replied "I daresay **you**
haven't had much practice. When I was your age, I always did it for
half-an-hour a day. Why, sometimes I've believed as many as six
impossible things before breakfast".
     This is from a children's fairy tale and you may well question its
relevance to the serious world of tomorrow. Let me then quote from a
leading British industrialist, Sir John Harvey Jones who, as Chairman
of ICI, demonstrated beyond doubt his ability to transform an ailing

giant chemical corporation into a powerful and profitable world force.
Giving advice to managers about planning for the future he wrote "It
has to be possible to dream and speak the unthinkable, for the only
thing we **do** know is that we shall **not** know what tomorrow's world will
be like.   It will be changed more than even the most outrageous
thinking is likely to encompass."

        We need to keep our imaginations intact, to perceive opportunities
and carry them through against the powerful forces of prejudice and be
will to accept what today may seem improbable if not impossible.   After
all, it took only a few short years for man to walk on the Moon after
President John F. Kennedy decreed that it would be done.

        In the context of the subject of this Congress I believe that
energy from the ocean will form part of this unthinkable future, but we
shall have to work hard to achieve it and I urge you to do so.

WAVE ENERGY TECHNOLOGY ASSESSMENT

GEORGE HAGERMAN[1]
TED HELLER[2]

1. INTRODUCTION

This paper presents a comparative survey of twelve wave energy
conversion devices. Table 1 identifies the developer of each
technology and indicates the largest scale at which each concept has
been physically tested. Note that eleven of these devices are designed
for electric power generation, and one (no. 12, DELBUOY) has been
developed for fresh water production. For the electric power
generating technologies in Table 1, "model" refers to a test where the
device's power conversion system was simulated, "prototype" refers to a
test where the device actually generated electricity that was used
locally or dissipated, and "demonstration" refers to a device that is
connected to a utility grid.

These twelve technologies were selected for this survey based on
evidence of long-standing program history, hardware testing in wave
tanks or natural waves, and the existence of commercial plant designs
for economic evaluation. These technologies also cover a fairly
complete range of alternatives in terms of wave energy absorption mode,
power conversion mechanism, working fluid, and type of platform
(floating or fixed). It is certain that the first generation of
commercial wave power plants will utilize at least one of these
technologies.

Space limitations prevent a detailed description of each device.
Some of the more important devices will be described in the oral
presentation of this paper. Additional information may be found in the
Reference section.

The first section of this paper presents a cost-of-energy
comparison among the electric power generating technologies. The
second section evaluates the economic potential of two heaving buoy
devices. This evaluation suggests that wave energy can be an
economically competitive source of both electricity and fresh water in

[1] SEASUN Power Systems, 124 East Rosemont Ave., Alexandria, Virginia
22301, USA
[2] Virginia Power, 5000 Dominion Blvd., Glen Allen, Virginia 23060, USA

D. A. Ardus and M. A. Champ (eds.), Ocean Resources, Vol. I, 183–189.
© 1990 Kluwer Academic Publishers. Printed in the Netherlands.

TABLE 1. Twelve near-term wave energy conversion technologies

No.	Device Name	Developing Company	Largest-Scale Physical Test
1.	KN System	B. Hojland Rasmussen A/S Copenhagen, Denmark	1 kWe prototype in the Oresund
2.	Hose Pump	Gotaverken Energy Gothenburg, Sweden	30 kWe prototype (1/3-scale buoys) off Vinga Island
3.	Wave Energy Module	NORDCO Ltd. * St. John's, Newfoundland Canada	1 kWe prototype on Lake Champlain
4.	Contouring Raft	Sea Energy Corporation New Orleans, Louisiana, US	1/15-scale model in wave tank
5.	Nodding Duck	University of Edinburgh Edinburgh, Scotland, UK	1/10-scale Model on Loch Ness
6.	Tandem Flap	Q Corporation Troy, Michigan, US	20 kWe prototype on Lake Michigan
7.	Neptune System	Wave Power International Sydney, New South Wales Australia	1/12-scale model in wave tank
8.	Multi-Resonant Oscillating Water Column	Kvaerner Brug A/S Oslo, Norway	500 kWe demonstration (full-scale plant) at Toftestallen
9.	Backward Bent Duct Buoy	Ryokuseisha Corporation Tokyo, Japan	1/10-scale prototype in the Sea of Japan and Mikawa Bay
10.	SEA Clam	Sea Energy Associates Ltd. & Coventry Polytechnic Coventry, England, UK	1/10-scale model on Loch Ness
11.	Tapered Channel	Norwave A/S Oslo, Norway	350 kWe demonstration (full-scale plant) at Toftestallen
12.	DELBUOY	ISTI-Delaware Lewes, Delaware, US	250 gpd prototype (full-scale buoy) off Magueyes Island

* Licensed by US Wave Energy Inc., Longmeadow, Massachusetts, US

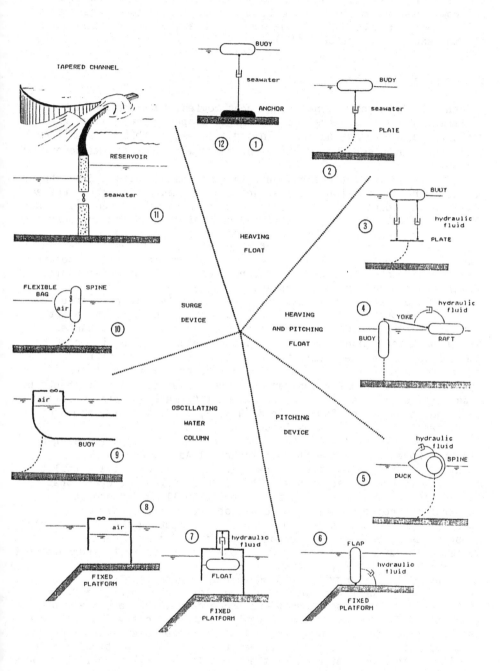

Figure 1.   Stylized representation of wave energy devices.

the Exclusive Economic Zone (EEZ), particularly for communities that
are now dependent on imported oil, and whose energy needs are not
sufficiently high to achieve the economy of scale required for ocean
thermal energy conversion (OTEC).

2.   COST OF ENERGY COMPARISON

For the electric power generating technologies, full-scale plant
designs that have been prepared by the developers were used as the
basis for economic evaluation. In order to establish a more uniform
basis for comparison, the developers' costs were modified as described
below.
      The largest plant designs involve hundreds of units and sea-to-
shore power transmission schemes incorporating multiple cable runs and
platforms dedicated solely to power conditioning and switching. Other
designs are intended for small island communities and involve only a
single submarine power cable. Therefore, the cost of power
transmission is more a function of plant size and distance offshore
than the particular wave energy device. Accordingly, this distorting
factor was removed in order to better reveal the relative economics of
the actual conversion technologies.
      Where capital cost breakdowns were available, absorber structure
typically accounted for over 50% of the cost of the offshore plant. It
was also evident that concrete and steel fabrication costs varied by a
factor of 2 to 3 among the various technologies. Again, in order to
more clearly reveal economic trends related to the efficient use of
structure, a uniform cost schedule was used that was thought to be
typical of small fabrication yards in the United States. It should be
noted that due to lack of cost breakdown information, this modification
could not be made for four devices (Nos. 3, 7, 8, and 11), and the
developers' original cost estimates were used in these cases.
      For each plant design, the cost of energy was computed using the
Electric Power Research Institute Technical Assessment Guide
(EPRI TAG, 1987). This formula reflects the cost of private utility
ownership in the United States. It should be noted, however, that the
first commercial wave power plants probably will be financed as
independent projects, where the owner/operator sells power to a
utility. There is no standard formula for computing the cost of energy
in such cases, because the methods of project financing are so many and
varied. Therefore, the EPRI TAG was used, since it is the formula most
familiar to US energy planners.
      The cost of energy calculated in this manner is plotted as a
function of incident wave power on a log-log scale in figure 2. Note
that for the two technologies where three data points are available
(Nos. 6 and 9), the logarithm of cost of energy closely fits a linear
relationship with the logarithm of incident wave power. It therefore
seems appropriate to visualize linear trends in this particular plot,
and it can be seen that most of the technologies fall on a line between
10 cents/kWh, at a resource level of 50 kW/m, and 20 cents/kWh, at a
resource level of 15 kW/m.

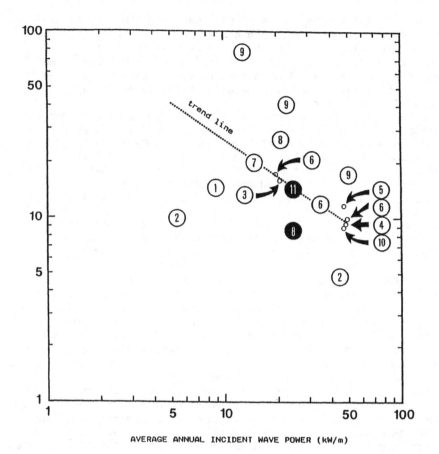

AVERAGE ANNUAL INCIDENT WAVE POWER (kW/m)

Figure 2. Cost of energy for eleven wave energy conversion devices
plotted as a function of annual average incident wave power. Dark
circles indicate pre-construction estimates for the demonstration
plants at Toftestallen, Norway. Note that much lower cost of
Gotaverken's hose pump device (No. 2).

It is important to note that figure 2 is a "snapshot" of eleven
technologies at very different stages of development. Some devices,
such as Ryokuseisha Corporation's BBDB are relatively new, and costs
may drop dramatically with further development. As with other emerging
renewable energy technologies, the economic picture for ocean wave
energy conversion is not a static one.

3.   WAVE ENERGY'S ECONOMIC POTENTIAL

An important finding of a recent study (SEASUN, 1988) for Virginia
Power and the North Carolina Alternative Energy Corporation is that
even with a modest wave energy resource of only 5.5 kW/m, the cost of

electricity landed ashore from a 25 MWe hose pump plant would be 9 to
15 cents/kWh (US), based on Virginia Power's method of calculating
revenue requirements. Although the intense wave energy levels (20-
30 kW/m) that occur in high latitudes (off northern California and New
Zealand, for example) have long attracted the interest of wave energy
developers, this recent finding suggests that wave power may be
economical in the tropics as well.

Furthermore, it should be noted that ISTI-Delaware's commercial
wave powered desalination system (No. 12, DELBUOY) is designed for
Caribbean islands and arid African coastal regions, where the wave
energy resource is typically less than 10 kW/m. At their prototype
test site off the southwest coast of Puerto Rico, they have
demonstrated continuous fresh water production at the rate of
250 gallons per day (950 liters per day) from a single buoy in
Tradewind-generated waves (1 m in height, 3-6 seconds in period). In
such a wave climate, they estimate the cost of water to be 19 to
22 dollars (US) per thousand gallons ($5-6/m^3) for a plant having a
capacity of 2 to 20 thousand gallons per day (8 to 80 cm^3/day).

Like ocean thermal energy conversion (OTEC), wave energy
conversion may be an economical source of both fresh water and
electricity for island and remote coastal communities that now rely on
diesel fuel. A recent cost-of-energy comparison between a 20 MWe land-
based open-cycle OTEC plant and a 25 MWe hose pump plant indicates that
levelized revenue requirements for the two technologies are nearly
identical (Hagerman and Heller, 1988). At smaller plant sizes,
however, the comparison is likely to favor wave energy conversion, due
to the modular nature of hose pump technology.

Unlike OTEC, wave energy is not a stand-alone, firm energy source,
nor should it be claimed as such. When used in conjunction with other
indigenous resources, however, such as marine biomass-derived fuel
(Gold and Schultz, 1986) and various solar technologies, it may be an
important component in a mix that provides energy self-sufficiently for
isolated communities whose power and water needs are not large enough
to benefit from the economy of scale required for profitable operation
of an OTEC plant.

The above comparison assumes a wave energy resource of 5.5 KW/m,
comparable to that found in the Baltic Sea. Although long-term
measured wave data from the tropical oceans is scarce, it appears that
wave conditions in Hawaii and elsewhere in the Pacific can be two to
three times more energetic than this, particularly along windward
coastlines.

This comparison also assumes no fresh water production by the wave
power plant. An intriguing possibility that has yet to be explored is
using wave energy for desalination and electric power production in a
single plant. Both DELBUOY and Gotaverken's hose pump use the large
ratio of buoy diameter to pump diameter as a means of amplifying wave-
induced dynamic pressures on the buoy hull. Resulting pump pressures
are as high as 4 to 5.5 MPa (580 to 800 psi) and are sufficient to
drive a Pelton turbine or develop reverse osmotic flow.

According to Hicks et al (1988), only 20% of the seawater from a
DELBUOY pump is passed through the osmotic membrane as fresh water.
The remaining 80% is discharged as high-pressure brine, which

conceivably could be passed through a turbine to generate electricity. The information presented herein suggests that the best features of these heaving buoy devices can be combined to develop a wave energy technology that is capable of producing both fresh water and electricity in modular units, at competitive costs, in the tropics and in high latitudes.

4.   REFERENCES

Electric Power Research Institute (1987) 'TAG^{TM} - Technical assessment guide, Vol. 3: Fundamentals and Methods, Supply-1986'.

Evans, D.V. and De O.Falcao, A.F. (Eds.) (1985) 'Hydrodynamics of Ocean Wave Energy Utilization', Proc. of IUTAM Symp., Lisbon, Portugal, Springer-Verlag (Berlin).

Gold, B.D. and Schultz Jr., E.B. (1986) 'Marine biofuel for rural coastal and island communities in developing nations', Energy for Rural and Island Communities, IV, Proc. 4th Int. Conf., Inverness, Scotland, Pergamon Press (Oxford), pp.87-92.

Hagerman, G. and Heller, F.P. (in press) 'Wave energy technology assessment for grid-connected utility applications', Proc. of RETSIE/IPEC 88, California (1988).

Hagerman, G. and Heller, F.P. (in press) 'Wave energy: A survey of twelve near-term technologies', Proc. Int. Renewable Energy Conf., Honolulu, Hawaii (1988).

Hicks, D.C., Pleass, C.M. and Mitcheson, G.R. (in press) 'DELBUOY: Wave-powered seawater desalination system', Proc. OCEANS 88, Baltimore, Maryland (1988).

McCormick, M.E. and Kim, Y.C. (Eds.) (1986) 'Utilization of Ocean Waves - Wave to Energy Conversion', Proc. of Int. Symp., La Jolla, California, American Soc. of Civil Engineers (New York).

SEASUN Power Systems (1988) 'Wave energy resource and technology assessment for coastal North Carolina', Report prepared for the North Carolina Alternative Energy Corporation.

OTEC DEVELOPMENTS OUT OF EUROPE

D.E. LENNARD
Ocean Thermal Energy Conversion Systems Ltd.
Orpington
Kent, BR6 0AY
U.K.

1.   INTRODUCTION AND BACKGROUND

The title of this paper is appropriate in two ways. "Developments out
of Europe" was chosen to describe what has been carried out or is
planned in Europe, by both national and international groups related to
ocean thermal energy conversion (OTEC). However, the term "out of
Europe" is appropriate too because Europe per se has no home market for
OTEC - the resource is not available. The resource is the ocean
thermal gradient (OTG), and it is suggested that this is what we should
be considering, rather than just OTEC.
      To set the scene the nature of the resource is briefly described
and, for convenience, is related just to energy.
      OTEC plants operate by extracting energy from the temperature
difference of warm surface waters and deep (approx. 1000 m) waters of
the oceans in the tropical and sub-tropical zones. They can be
floating or land based. This free-fuel system can operate with open or
closed cycles, with an overall plant efficiency of little more than
2.5% for temperature differences of 20°C between the warm and cold
waters. In the case of open cycle systems the warm water is flash
evaporated at low pressure, passes through a turbine, and is condensed
by the cold water. For the closed cycle a working fluid (such as
ammonia) is vaporised in the warm water heat exchanger, passed through
a turbine, and condensed in the cold water heat exchanger. The
turbines drive a generator.
      In addition or as an alternative to power generation, OTEC plants
can be used to produce potable water (from flash distillation for
example) or for fish farming (using the cold deep nutrient rich water),
as well as for other processes. Whilst technically feasible,
acceptable economy of some of these processes is not proven.
      OTEC is an environmentally benign power source for the quantities
presently being considered and unlike other renewables is a base load
system.
      The theoretical size of the resource in the tropics is to all
intents and purposes infinite. However, the total output that would be
economic at present is probably only a few hundreds of MW, being the
demand for small plants in relatively remote islands where oil

191

generated power is expensive. Many developing countries meet the criteria for OTEC power to be economically viable in due course.

As an "EEZ Resource: Technology Assessment" (the title of this Conference), OTEC is unusual in that it inherently produces a variety of "products". Other marine renewable energies (wave, tidal for example) produce just that - renewable energy (which can then be used for a variety of activities). OTEC certainly produces energy; but the potable water, and the nutrient rich deep water resource, result in two other complete scenarios of output - and both are of substantial interest in their own right in many of the developing countries of the tropical and sub-tropical zones where the OTEC resource occurs.

To the Europeans (with no home market for OTEC) the diversity of product line is of considerable appeal - spreading the risk of the investment. Indeed, another paper at this conference* illustrates this point very well.

2.   EUROPEAN ACTIVITY

As with countries outside Europe, considerable impetus was given to European interest in OTEC with the large price increases in oil - although France in particular had been working on the technology in the 1960s - as befits the nation which produced d´Arsonval the "founder" of the theory over 100 years ago, and Claude, an early developer of OTEC over 50 years ago.

The current low price of oil has undoubtedly lessened the OTEC activity (out of Europe) but the teams remain in place, and progress continues - albeit at a reduced rate. However, this slow down has provided the opportunity to fine tune some of the component items - for both energy and non-energy applications - the result of which should be engineering solutions closer to the optimum.

Activity is described below, by nation in alphabetical order, followed by international initiatives relevant to Europe.

2.1 Belgium

During the last decade interest has been in the aquaculture aspects of OTEC and not the energy generating part of the system, and involvement in the choice of sites to minimise environmental impact. Initially this was through the Eurocean (see below) programme, and more recently has been in connection with the proposed Indian OTEC activity.

2.2 Finland

The Finnish involvement has been with the Swedish activity, and concerned particularly with contributing desalination equipment which operates on the small temperature differences of OTEC systems. The involvement was initially as an extension of the European work on OTEC, with desalination and aquaculture, and then as a part of a 1 MW programme proposal for Jamaica. In both cases the plant concepts were

* by Dr. F.A. Johnson, G.E.C. (UK)

of the land based type.

2.3 France

From the 1880s and 1930s activity, France had a lead, and re-opened its
current round of interest with a programme commencing in 1978. A
feasibility study by CNExO and French industry examined both floating
and shore based designs, with open and closed thermodynamic cycles.
OTEC plants up to 20 MW looked technically feasible, and at 10 MW
looked economically interesting with oil at 1982 prices.

The French Pacific island of Tahiti is where the French programme
will be carried out. Initially, two industrial groups were involved,
now fused into one. Designs for a 5 MW land based plant in Tahiti
continue, and a decision on closed or open cycle, based on experiments,
remains to be taken. The experiments covered heat exchanger systems
for the closed cycle, evaporators and condensers for the open cycle,
together with material evaluations for the cold water pipe and detailed
site surveys in Tahiti. Co-ordination and partial funding of the
programme was undertaken by IFREMER, which has succeeded CNExO and
ISTPM, and original plans called for prototype operation by 1990.
However, from 1986 work has concentrated on refining the economic
analysis and a valuable report summarises French views on: cost
comparisons for power generation between OTEC and diesel power (in the
range from a few MW to some tens of MW); the cost of producing potable
water via OTEC when compared with conventional desalination techniques
(from a few thousand m^3/day to a quarter of a million m^3/day); and
finally the benefit of producing fresh water as a direct by-product of
an open cycle OTEC power plant.

Further progress of the French programme is dependent on oil
prices moving to a level which will show benefit from OTEC developments
according to these economic comparisons.

2.4 The Netherlands

In addition to their own programme, aimed initially at Curaçao, in the
Netherlands Antilles, the Netherlands were also involved in the
international Eurocean programme from the beginning.

Both floating and land based options were considered for Curacao,
related to electricity production and desalination. A later closed
cycle land based design for Bali was for a 0.25 MW size plant and plans
for this began in 1982.

With surface water temperatures in excess of 29°C for part of the
year, the required 20°C difference was obtainable with a cold water
pipe reaching to barely 500 m depth. This is a significant benefit,
taking note of the difficulty of construction, and cost, for cold water
pipes of both land based and floating plants.

The design of cold water pipe proposed has moved on from a
membrane stretched over the light-weight concrete rings to the work for
Bali, where the proposal is to use a high density polythene pipe,
smooth on the inside but with an external wall shaped to incorporate a
hollow spiral profile. This design provides a capability to resist

external pressure differentials whilst permitting adequate bending
flexibility.

For the power cycle ammonia has been chosen in preference to
freon, since heat transfer coefficients are better, and particularly
since a falling-film heat exchanger is preferred. Bio-fouling control
is considered effective using intermittent chlorination at dosages
which, it is believed, will not have harmful environmental effects,
although in common with all OTEC proposals this aspect is recognised as
the one which needs careful long term evaluation in a system which is
otherwise environmentally benign.

Heat exchanger material choice is considered to lie between
aluminium and titanium; whilst aluminium has cost in its favour,
"state-of-the-art" preference applies to material choice, and for early
installations titanium is the Netherlands preference.

The scheme for Bali is, at the time of writing, on hold.

2.5 Norway

Following a small involvement in Caribbean OTEC activities, there
emerged in 1986 the Norwegian proposal for ROTEC - or Rock OTEC -
proposed by a group of Norwegian industrial interests. The scheme owes
much to the experience gained from Norwegian rock based hydro-electric
plants, and in particular removes the ´conventional´ cold water pipe of
a land based OTEC plant since the cold resource is drawn up through an
intake drilled in the rock. Instead of the more usual ammonia or
freon, carbon dioxide is the working fluid, and a more or less standard
Francis turbine is used to extract power. Locations suitable for ROTEC
require a steeply sloping sea bed off the coast in order to make the
cold water intake a feasible and economic component. Enquiries have
not indicated whether any practical experimentation and development has
been done, or whether any sites for installation have been located or
selected.

2.6 Sweden

Like the Netherlands, Sweden developed its own programme out of that
initiated with Eurocean. From 1980, work developed on closed cycle
land based OTEC prototypes, and in the spring of 1983 a project scheme
for a 1 MW plant in Jamaica was initiated. It is noteworthy that the
location, as for the Netherlands programme, was an island site with oil
generated electricity. The Jamaican project involves both a design
engineering organisation and a heat exchanger manufacturer. Between
the Eurocean and Jamaican work, Swedish interests examined the
prospects for production of fresh water and food from an OTEC plant, in
addition to electrical power and this was done in co-operation with
Finnish interests referred to above.

Considerable practical testing of bio-fouling on heat exchangers
has been carried out in various oceans of the world.

2.7 United Kingdom

The British programme also formally came into being in 1980, although
preliminary work had been going on for half a decade. It consists of
an industrial lead, but involves substantial activity from two
universities, and some financial support has also been provided by the
Department for Enterprise (DTI).
    Work has concentrated on development of a floating 10 MW closed
cycle demonstration plant, following on from an evaluation of likely
sites, and the state-of-the-art in technological developments. During
1983 the university groups started programmes of related work dealing
with the dynamic response of the system, aspects of design for the
heat exchangers, and features of economic and risk assessment of OTEC
systems, and this was developed for a further two years to 1987. The
programme is directly relevant to, and planned in co-ordination with,
the industrial Company's work noted above.
    The preferred conceptual design has been selected, although work
continues on potentially encouraging alternatives, and three sites (all
islands) are being evaluated - initially in the Caribbean but now the
Pacific is preferred.
    The philosophy of the design is to use standard state-of-the-art
items where possible, but the cold water pipe has to be substantially
new technology. It is the view of the UK team that whilst the problems
of cold water pipe for a floating plant are different from those of a
land based or shelf based design they are no more difficult, and in
some senses are less difficult, to design. Whereas this view is not,
in general, held by teams in other countries, the development of
tension leg platform design for the offshore oil and gas industry in
deeper waters has further encouraged the UK team in their views.

2.8 USSR

Little is known about the USSR work on "reverse OTEC" developments,
with a view to using the sea off the coast of Siberia as the warm
resource and the overlying air as the cold resource for an OTEC cycle.
Clearly, the quantity of air which must be pumped will be very large
indeed to achieve an equivalent thermal mass to the water used in
"conventional" OTEC plants. No published documentation describing this
work is known to be available, and requests for information following
the Russian wish to participate in a study by the World Energy
Conference in 1986 have still not produced any response. The "reverse"
OTEC concept is an interesting technical exercise, and the
international OTEC community at large would be very interested to hear
details of the work undertaken, and the results obtained - in
particular as far as air movement is concerned, and the overall
economic calculations.

3. MULTI-NATIONAL INITIATIVES

3.1 Eurocean

An international R&D organisation - the European Oceanic Association
(Eurocean) based in Monaco, undertook extensive paper studies on OTEC
and ODA (OTEC-desalination-aquaculture) during the period 1977-81. A
general study on marine renewable energies was followed first by an
outline design for a nominal 10 MW floating plant. After this 9 out of
the 30 member companies, drawn from 5 of the 8 nations in Eurocean,
carried out a more detailed study for a combined OTEC, desalination and
aquaculture plant (1 MW nominal size) and also for a rather larger
floating plant to generate electricity only. A number of companies
have gone further forward with some of the national plants mentioned
above. No further OTEC activities are planned by Eurocean itself at
present, but schemes are in hand for a further review of priorities in
1988.

3.2 Commission of the European Communities (CEC)

In 1983 the CEC (Directorate General XVII - Energy) commissioned
reports on wave, tidal, OTEC and offshore wind power and in 1984 the
reports were published. The work on OTEC had been undertaken by the
Chef du Service Energie of the French organisation IFREMER, who
consulted widely in preparing his report. Among the conclusions was a
firm recommendation to construct a sufficiently large demonstration
plant - 5-10 MW was proposed. As well as reviewing past international
progress, the report also proposed the topics for future detailed R&D
activities, leading to the construction of the demonstration plant. Up
to the present (June 1988) OTEC has not been included in the CEC
renewable energy programme, which is aimed primarily at "Energy for
Europe", and as noted at the beginning Europe is unsuitable for OTEC in
lacking a resource. Recently though (April 1988) Directorate General
XII - Science, Research and Development - has initiated plans involving
marine science and technology and it is understood a number of
proposals related to OTEC have been received. There is much logic in
this when the very substantial CEC programme from Directorate General
VIII (Development) is considered, which provides funding for the Lome
programme - beneficiaries being 66 nations, most of whom are in the
"OTEC belt" - or should it be called the OTG belt (see below).
     Let us dwell further on the Commission of the European Communities
interest in this topic. In the Introduction to this paper the broader
opportunities resulting from the thermal gradient in the oceans have
been described. There is drinking water; there is aquaculture; there
are many other potential products. Figures 1 and 2 give examples.
Should we therefore be speaking now only about ocean thermal energy, or
all the potential outputs derived from the ocean thermal gradient: in
fact OTG, rather than OTEC alone? The interrelations between OTG and
the CEC have been indicated above as far as energy alone is concerned.
But if potable water and mariculture are added (as they probably should
be) the relevance of OTG to the interests of the CEC and the countries
of the EC increases dramatically. The DGVIII interest has been cited -

which relates to technology transfer, and development of the resource of the Lome countries. But perhaps, at this stage of OTG applications, it is the Marine Science and Technology (MAST) programme of DGXII (also mentioned above) which is most relevant. It is concerned with "precompetitive research", and it is this which fits exactly the opening up of opportunities for OTG - of interest to many countries of the EC. Priorities for the MAST programme are presently being set. OTG is right from Lome countries, and right for the EC countries; perhaps then the title of this paper should be "OTG Applications out of Europe".

3.3 United Nations

It would be wrong to conclude this nation by nation resume without reference to the UN - since European countries are members, and the UN has examined the topics of OTEC. In 1981 the UN Conference on New & Renewable Sources of Energy was held in Nairobi. One of papers prepared for that meeting - its authors including the major participants in OTEC at this time as well as others concerned with this technology - painted a reasonably optimistic picture for the future of OTEC in reviewing various forms of ocean energy. A draft programme of action for the UN following that meeting included further work on OTEC and in 1984 the UN published "A Guide to Ocean Thermal Energy Conversion for Developing Countries", which reviewed the technology, the economics, and the particular opportunities which were seen for developing island nations in particular.

4.   CONCLUSIONS

The last line above indicates where the opportunities lie. This paper is not called upon to review the "pro's and con's" of OTEC, merely the European position. However, that position depends fundamentally on the market being established (as the Introduction indicates), and the economy proven. These two are interdependent. As the economy improves, so does the market, and so in turn does the attraction of investment by the financial sector in OTG opportunities. A view expressed by many Europeans concerned with OTEC is that the next step has to be construction of a representative size of demonstration plant. Such a plant would be a very suitable topic for a European consortium - and the current initiative of the CEC appears to be a timely potential complement. Islands are undoubtedly the best place for OTEC applications - and a number of Lome island countries are very attractive for the resource, and for the market (energy, potable water, aquaculture). In the short term therefore, there is a strong case for support of OTG research through, in particular, the MAST programme of DXGII of the CEC. To complement that the OTEC part of the resource is ripe for demonstration, and other parts of the CEC have a real role to play here, as described in the paper.

5.   ACKNOWLEDGMENT

I would like to thank those who have provided information to me in the preparation of this paper - in particular colleagues from France, The Netherlands and Sweden.

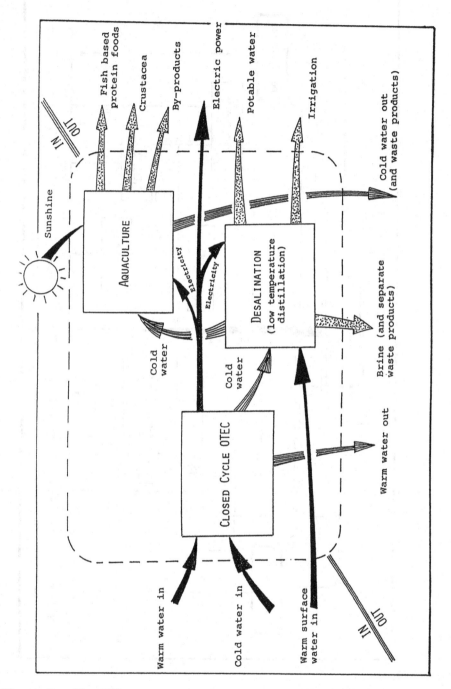

Figure 1.   The OTG resource - schematic arrangement of combined OTEC,
desalination and aquaculture plant (after Eurocean).

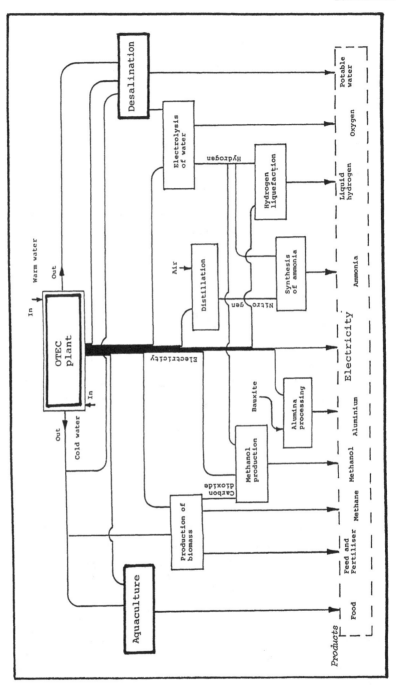

Figure 2.   A range of products which can be derived from a plant making use of the UTG resource (after Eurocean).

ENERGY FROM THE OCEANS: A SMALL LAND BASED OCEAN THERMAL ENERGY PLANT

DR. F.A. JOHNSON
Technical Director, GEC-Marconi Research Centre
Great Baddow, Chelmsford, Essex, CM2 8HN
England
U.K.

ABSTRACT. This paper describes a small land based closed cycle Ocean Thermal Energy Plant which is being designed to supply deep nutrient rich sea water to an aquaculture facility and to produce a net electric power output of up to 300 kW. In addition the plant will also demonstrate a small air condenser for producing potable water at a rate of 2650 US gallons per day. The work is being done for ALCAN International and the plan is to build and operate such a plant at Keahole Point in Hawaii.

1.   INTRODUCTION

The greatest resource in the ocean is the water itself. The sun's radiation is very rapidly absorbed in the surface layers of the oceans thus generating a huge reservoir of warm water and huge quantities of water vapour. However, the sun's radiation does not penetrate to great depths and this, coupled with cold polar currents, keeps the vast mass of deep water at temperatures of 4°C or less. In tropical areas the temperature difference between surface and deep water is typically 20°C or more.
    The possibility of utilising the temperature difference between surface and deep sea water to drive a heat engine has been recognized for a very long time. However, it has also been recognized that with the relatively small temperature differences involved one would need to process a very large amount of heat energy to extract any useful work. Thus we have the challenge - can we utilise the virtually unlimited resources of warm and cold sea water to generate electrical power?
    The first experiments to answer this question were carried out in Hawaii in 1979-80 where a floating heat engine was made that generated a gross electrical output of 50 kW. However, nearly 80% of this power was needed to pump the sea water through the heat exchangers and the working fluid from the condenser to the evaporator so that the nett output was about 10 kW. Experiments with this plant were conducted for about 3 months and they demonstrated that a positive nett power output was possible.

D. A. Ardus and M. A. Champ (eds.), Ocean Resources, Vol. I, 201–211.
© 1990 *GEC-Marconi Ltd.*

Shortly after this the Tokyo Power Services Company designed and built a larger land based experimental plant on the island of Nauru. This plant produced a gross electrical output of 120 kW and a nett output of some 31 kW. This plant was operational over a period of 9 months during 1981-82, and achieved a maximum continuous period of operation of 10 days. Subsequently the cold water pipeline was destroyed by a typhoon.

These two experimental projects stimulated a number of design studies and laid the foundations for further work on materials and bio-fouling. The design studies indicated that if an OTEC plant was to produce electricity in a cost competitive way with that of the most advanced fossil fuel plants the nett output would need to be increased from 31 kW to 10 MW or more, the availability increased to about 80% or better and the plant operated for about 25 years or more. While this is a feasible objective it is a very large extrapolation from proven technology and consequently there is a high risk of meeting unforeseen problems which can prove to be very costly.

From a technical point of view what is needed, I believe, is a more modest scaling up of the previous Japanese work. A closed cycle land based plant with a gross output of 400 to 700 kW connected to a distribution grid and operating for a minimum of 5 years with a high availability would provide a very thorough test of the whole OTEC concept and in particular the materials and fabrication techniques. The sea water pipes would need to be reliable, the sea water pumps would need some redundancy, control systems would need to be installed to match the output to the distribution grid, maintenance cycles would need to be planned and, in addition, the plant would need a monitoring package to check the correct working of all the key components. Such a demonstration would provide a sound basis on which to plan, design and finance larger OTEC installations.

However, as the design studies have shown, a plant of this type would be doing well to produce an annual return on capital of 1%, scarcely enough to pay the staff and not the sort of proposition to appeal to managing directors or to venture capitalists. This highlights the dilemma for OTEC namely:

(i)   the economics of power generation suggests that a very large leap forward is needed;

(ii) engineering prudence suggests that an incremental approach is the soundest way to proceed;

(iii) the problems of financing an incremental approach are complicated by the fact that as electric power plants they are uneconomic;

(iv) the problems of financing a large leap forward are bound to be considerable in view of the risks of meeting unforeseen problems.

These problems are a natural corollary of the second law of thermodynamics which seem to indicate that "you can't win". The question is thus - can one break out of this situation?

2. A SMALL LAND BASED PLANT

My own view of OTEC has been considerably modified by the work done at
the Natural Energy Laboratory of Hawaii. This had measured
consistently high levels of nitrates and phosphates in sea water pumped
up from the deep ocean around Hawaii. These inorganic nutrients are
essential for plant and hence marine growth. The major fisheries in
the oceans are in areas where natural upwellings of this deep cold sea
water occurs. Further this deep sea water proved to be remarkably free
from pathogens so was ideal for aquaculture. Subsequent work at this
laboratory had demonstrated the value of this deep ocean sea water
resource for aquaculture and two commercial enterprises have already
started there. In my view, this work had demonstrated that the
greatest resource in the oceans is the water itself.
     The combination of aquaculture with a small scale OTEC plant
should dramatically alter the economics of such a combined operation.
The economics of aquaculture is such that it can support the costs of
installing a cold water pipe and its associated sea water pumps. If
one views the OTEC plant as just the warm water pipe, heat exchangers
and power generation equipment, not only is the capital investment much
smaller but the OTEC plant can supply all the pumping power required
plus a useful surplus of electricity for sale. This suggests that such
a composite plant would be a commercial possibility and thus provide
the means for an incremental development of the OTEC concept itself.
     This view was arrived at, quite independently, by ALCAN
International, in particular by Nigel Fitzpatrick. As a result of a
number of discussions ALCAN asked us if we would design such an OTEC
plant for them and this we are now doing. The design aim is a plant
that could provide a minimum of 13,000 US gallons of deep ocean sea
water per minute for aquaculture and provide a nett output of 150 to
300 kW of electricity into a local electricity supply grid. Our
preferred site is close to the Natural Energy Laboratory at Keahole
Point on the Big Island of Hawaii. This not only has excellent access
to deep ocean water, a local grid system willing to buy all surplus
electricity, a growing aquaculture industry based on pumped ocean water
but also all the infra-structure needed for installing and maintaining
the plant. Of even greater importance is the interest of senior
officials in the State of Hawaii in such a project.

2.1 Design Considerations

The design work falls into two distinct parts. The initial work has
been concerned with studying the various parameters such as flow rates,
pipe sizes, heat exchanger types, working fluids, fabrication
techniques, etc., with a view to establishing a minimum cost option.
This will then be followed by the detailed design of this chosen
option. In the course of this detailed design there may well be some
further changes in the key parameters as specific areas are
investigated in more detail. The main part of this design work is
being done in our Engineering Research Centre by a team headed by Dr.
G.J. Williams.

2.2 Sea Water Pipes

The largest single expense in such a small land based OTEC plant is the
sea water pipelines and the sea water pumps. Preliminary estimates
indicated that we would probably need pipes of about 36 to 40 inches in
diameter to cope with the flow rates without the need for excessive
pumping power. We sub-contracted this part of the work to Makai Ocean
Engineering Inc. at Waimanalo, Hawaii. They have considerable
experience in the design and installation of sea water pipelines in
Hawaii. They have designed three sea water pipelines of 12, 18 and 40
inches that have been installed at Keahole Point. All of these
pipelines use polyethylene as the pipe material. Their preliminary
work indicates that we shall need a cold water pipeline of some 2300 to
2500 m in length to reach down to a depth of 640 m. The warm water
pipeline will be about 300 to 380 m long to reach an intake depth of
30 m at a point some 10 m above the bottom - this is to minimise bio-
fouling problems. There can also be problems with disposing of the
discarded sea water from both the OTEC plant and the aquaculture
facility. Here we are considering mixing these two outputs and
discharging them through a sea water pipeline at a depth just below the
mixing layer, i.e. at a depth of about 100 m. To do this will require
a third pipeline of approximately 850 to 900 m in length. The greatest
part of the cost of these sea water pipelines is the transition section
from the shore to a point sufficiently deep to be free of surface wave
loading.
        Following Makai's recommendations we plan to site the sea water
pumps at a depth of about 10 m and use three pumps for each pipeline,
two operating in parallel and the third as a standby. Stainless steel
pumps are preferred but they are considerably more expensive than cast
iron pumps. However, they have a considerably longer life. These sea
water pumps consume a significant proportion of the gross power output.
Further, as the exhaust cold water is to be delivered to an adjacent
aquaculture site one cannot use a siphon to reduce the pumping power
needed. At Keahole Point the land is about 3 m above mean sea level so
this is a further important factor. Another consideration is that the
cold water supply to the aquaculture facility will need to be
maintained during any plant shut-down for maintenance so suitable by-
pass circuits need to be provided. These considerations lead to a
general layout for the plant as shown in figure 1.

2.3 Heat Exchangers

The key components in a closed cycle OTEC plant are the heat exchangers
used in the evaporator and the condenser. Our initial designs
considered conventional heat exchangers. Conventional shell and tube
heat exchanger are the most expensive to fabricate but have moderate
pressure drops and thus moderate pumping losses. Conventional plate
and fin heat exchangers are considerably less expensive to fabricate
but, to achieve good heat transfer characteristics, they use high fluid
velocities which require considerable pressure drops and thus
considerable pumping losses. Our conclusion was that neither were
optimised for OTEC applications.

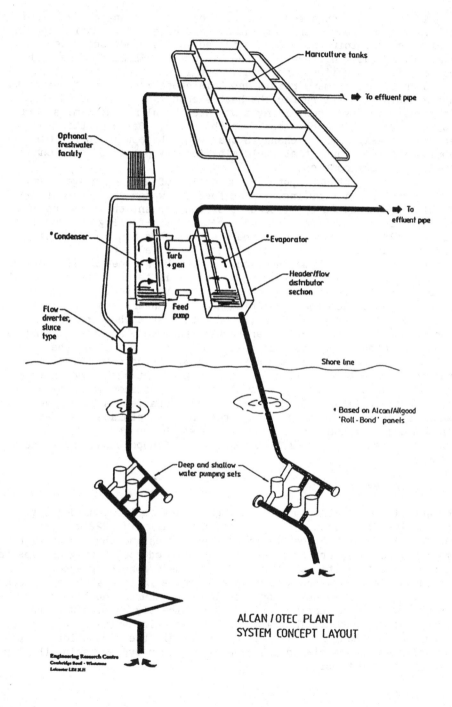

Mariculture tanks

To effluent pipe

Optional freshwater facility

To effluent pipe

*Condenser

*Evaporator

Turb + gen

Header/flow distributor section

Flow diverter, sluice type

Feed pump

Shore line

* Based on Alcan/Allgood 'Roll-Bond' panels

Deep and shallow water pumping sets

ALCAN / OTEC PLANT
SYSTEM CONCEPT LAYOUT

Engineering Research Centre
Cambridge Road - Whetstone
Leicester LE2 2LH

Figure 1.

An important input for the design of heat exchangers came from
ALCAN. They had been carrying out a long series of corrosion tests on
a number of aluminium alloys in both deep and surface sea water - the
chemistry is quite different in the two cases. In surface sea water
the corrosion is uniform and regular but in deep sea water pitting
corrosion is the primary result and this needs more care if one is to
have a satisfactory life for the heat exchangers. The introduction of
a thin surface layer containing zinc into the aluminium limits this
pitting corrosion to the zinc layer thickness. Based on this work we
decided to design heat exchangers that would be optimised for OTEC and
fabrication techniques that would minimise any possible local changes
in the alloy composition.

Another important input for the design of heat exchangers was
ALCAN's work on bio-fouling. In surface sea water regular small doses
of chlorine, produced by electrolysis, can prevent the build up of bio-
fouling. If the electrolysers malfunction, bio-fouling builds up
quickly. After maintenance to the electrolysers the bio-fouling
largely disappears but the thermal resistance falls to a value of 10%
higher than before the malfunction. It can be reduced to its original
value only by mechanical cleaning so this will be an important aspect
for both the design and maintenance of the evaporator.

We have now designed aluminium heat exchangers whose pressure
drops are less than that of shell and tube heat exchangers and whose
fabrication costs are lower than that of the plate and fin exchangers.
These heat exchangers will use roll bonding techniques for the main
part of their fabrication and either adhesive bonding or aluminium
brazing for a few specialised sections. Our heat exchangers have also
been designed to be mechanically cleaned and replaced in sections while
the OTEC plant is operating. This not only enables us to remove
sections for inspections but also enables up to replace them, at
minimum cost, without needing to shut down the whole plant. This
modular design will also make it easier to transport and erect the heat
exchangers at the site.

2.4 Power Generation

The actual form of the turbines and gearboxes varies considerably with
the working fluid used. We have considered ammonia, R22 and iso-
butane. Ammonia produces the highest pressures and thus match with
small high speed turbines which would then need reduction gearboxes to
connect them to suitable generators. The need for the gearbox
increases the losses in converting from mechanical to electrical power
output. R22 and iso-butane produce lower pressures and thus match into
larger lower speed turbines which can be directly coupled with the
electrical generator. They also ease the requirements on the strength
of the heat exchangers. We will carry out all the regulation and power
conditioning on the electrical side rather than by regulating the input
pressure to the turbine.

2.5 Optimisation

There are such a large number of parameters to consider in the design
of a closed cycle OTEC plant that the problem of optimisation can
become very complex. We found that the best way to handle the
optimisation problem was first to fix the following parameters:

(i)   the length of the sea water pipes and the intake temperatures -
      these are fixed by the local conditions;

(ii) the height above mean sea level at which the discharge of the sea
      water takes place - again fixed by local conditions;

(iii) the water flow velocities and spacings in the evaporator and
      condenser that define the heat transfer rates.

The next step was to choose the following parameters:

(i)   a common diameter for the two sea water pipes;

(ii) the working fluid;

(iii) the sea water exhaust temperatures from the evaporator and the
      condenser;

(iv) the working fluid input temperature to and exhaust temperature
      from the turbine - note that the working fluid temperatures must
      lie within the range between the sea water exhaust temperatures
      from the evaporator and the condenser;

(v)   the turbine mechanical shaft power.

From the above data one can calculate the evaporation and condensation
rates of the working fluid and, of course, the power needed to feed the
condensed working fluid back from the condenser to the evaporator.
From these results one can then scale the size of the two heat
exchangers to handle the thermal loads and derive the sea water pumping
requirements needed. The next step is to calculate the nett electrical
power that would be generated by making allowances for the following:

(i)   the losses in converting the shaft power to electrical power;

(ii) the pumping power needed to produce the sea water flow rates in
      the evaporator and condenser;

(iii) the pumping power needed by the feed pump that returns the working
      fluid from the condenser to the evaporator.

These calculations were carried out for a range of turbine shaft powers
to produce a graph of the nett electrical power available as a function
of the turbine mechanical shaft power.

At low shaft powers the nett electrical power becomes negative largely because of the power needed to raise the sea water above sea level. At high shaft powers the nett electrical power also becomes negative due to the fact that the pumping losses, for fixed sized sea water pipes, eventually increase more rapidly than the gross shaft power. Thus these curves always show a maximum nett electrical output power for some particular turbine shaft power. Not surprisingly this maximum nett electrical power available, and the corresponding turbine shaft power, increase with the diameter of the sea water pipes.

This analysis demonstrated, very clearly, the comparison between different working fluids. We found that R22 can produce the greatest maximum nett electrical power, although the results for ammonia are only slightly less. This is because with R22 the turbine shaft rotation rate is suitable for direct connection to the electrical generator so the conversion loss is smaller. In the case of ammonia the turbine shaft rate is over four times greater so a gearbox is needed between the turbine and the generator which results in a considerably higher conversion loss which more than offsets the slightly lower pumping power requirements. However, both R22 and ammonia can produce nearly 80% more power than iso-butane. This is because they have more favourable heat functions (enthalpies) than iso-butane. In figures 2, 3 and 4 we show some early results demonstrating this point. Another interesting feature that this analysis has shown is that when one moves from summer to winter conditions the nett electrical power falls significantly but, by re-optimising the thermal transfer rates in the evaporator and condenser, this loss in nett electrical power can be minimised. An early example of this is shown in figure 5.

2.6 Other Considerations

Care will be needed in the design of the safety and protection equipment and procedures will need to be established for the normal start-up and shut-down of the plant. For example the cold water pipeline will contain some 2000 tonnes of sea water which will need to be accelerated to a velocity of about 1 m per second during the start-up process - not something that can happen in a few seconds. Even the warm water pipeline will contain some 300 tonnes of sea water which will also need to be accelerated to a velocity of about 1 m per second. We will also need to have measures to cope with any forseeable malfunction of the equipment.

Another important aspect of this design will be the provision of a reasonable comprehensive instrumentation package. It will be important to measure not only the initial performance in order to compare it with the designed values but also to detect any slow drift in this performance away from the initial values and make an early assessment of the cause. From this we would expect to modify our maintenance schedules so that the performance could be held within acceptable bands. One of the key parameters that we shall be monitoring is the heat transfer rates in the heat exchangers. This instrumentation package will enable us to accumulate a great deal of new data about the actual operation of a closed cycle OTEC plant.

2.7 Water Vapour Condensers

There is one additional item we are proposing to demonstrate, namely
the generation of potable water. Any shore based OTEC site will have a
virtually unlimited supply of warm air with a dew point well above the
exhaust temperature of the cold water discharge from the OTEC plant.
allowing for the latent heat of water vapour and assuming a 30%
efficiency for an air condenser, one should be able to condense potable
water at a rate of 0.043% of the cold water flow rate for each 1°C
increase in the cold water temperature. This may sound very small but
OTEC is all about exploiting small efficiencies. In our case we could
increase the cold water exhaust temperatures from the condenser to 14°C
which is suitable for the aquaculture facility. This would suggest
that we could condense potable water at a rate of about 31,500 US
gallons per day. We are currently planning a small scale test facility
to produce somewhere in the region of 2650 US gallons per day and this
part of the work has been subcontracted to Dr. P.F. Monaghan of the
Department of Mechanical Engineering, University College, Galway,
Republic of Ireland. He has been doing much design and testing work on
external air heat exchangers for use with heat pumps.

3.   SUMMARY

We believe our design work has made a significant advance in the design
of aluminium heat exchangers for closed cycle OTEC plants. These heat
exchangers have a lower cost and lower pumping losses than shell and
tube or plate and fin designs. We also plan to make further corrosion
tests on sample heat exchanger components. In addition we have
designed an efficient water vapour condenser that can add to the
utility of a closed cycle OTEC plant. The hope is to complete the
detailed design work early in 1989. If all goes well we should then
begin to start construction work on this OTEC plant which should
demonstrate, in a very realistic way, the overall capability of the
closed cycle OTEC concept.

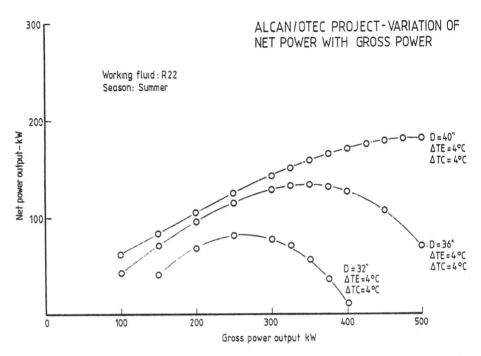

Figure 2.

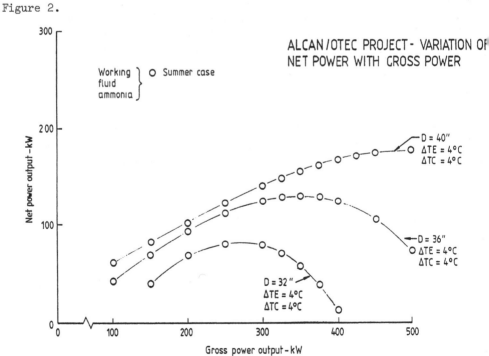

Figure 3.

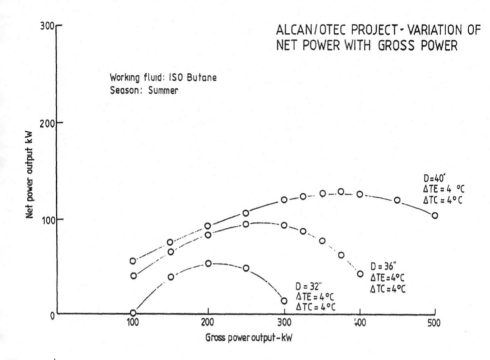

Figure 4.

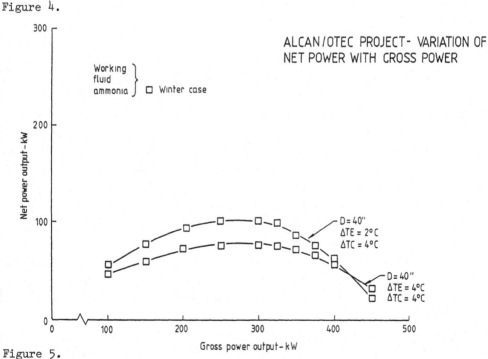

Figure 5.

A STATUS ASSESSMENT OF OTEC TECHNOLOGY

L.J. ROGERS[1]
R.J. HAYS[2]
A.R. TRENKA[3]
L.A. VEGA[2]

1, INTRODUCTION

The following discussion presents a status of the technology, market, industry, and economic viability of Ocean Thermal Energy Conversion (OTEC) systems. The main focus is on the Open-Cycle OTEC (OC-OTEC) investigations being conducted primarily under the auspices of the Department of Energy (DOE) Ocean Energy Program (OEP). A summary of technology advances which have enhanced the economic prospects for both open- and closed-cycle OTEC are presented. Of key importance was the demonstration of the technical feasibility of using the OTEC flash-evaporation of seawater process to produce fresh water. The production of such "by-products" as fresh water from the OTEC cycle significantly improves its cost effectiveness in producing electrical energy. A synopsis of planned activities which will culminate in a test facility designed to establish the technical feasibility of producing net power from the OC-OTEC system is presented. The facility will be used to conduct the Net Power-Producing Experiment (NPPE).

2. DEFINITION OF TECHNOLOGY

Recent advances have been made in the technology associated with Ocean Thermal Energy Conversion (OTEC). These advances are significantly increasing the attractiveness of OTEC while, at the same time, reducing the overall technical risks associated with bringing OTEC systems to the marketplace.

[1] US Department of Energy, Washington DC
[2] Pacific International Center for High Technology Research, Honolulu, Hawaii
[3] Solar Energy Research Institute, Golden, Colorado

D. A. Ardus and M. A. Champ (eds.), Ocean Resources, Vol. I, 213–231.
© 1990 *Kluwer Academic Publishers. Printed in the Netherlands.*

OTEC systems utilize the temperature difference between the cold, deep seawater from depths of 600 m to 900 m and the sun warmed surface waters to generate electricity. There are two types of OTEC cycles which can be used to convert the thermal energy to electrical energy. They are: Closed-Cycle OTEC (CC-OTEC) and Open-Cycle OTEC (OC-OTEC). There are also variations and combinations of these two basic cycles, namely "Hybrid Cycle" and "Mist Lift". In this paper only OC and CC will be addressed.

In the CC-OTEC system, warm seawater (25°C) is used to vaporize a working fluid such as freon or ammonia through a heat exchanger (evaporator). The vapor expands under moderate pressures, turning a turbine attached to a generator to produce electricity. The vapor, upon exiting the turbine, is recondensed using cold seawater (5°C) in another heat exchanger (condenser). The working fluid remains in a "closed" system and is continuously reused.

In the OC-OTEC, warm seawater is evaporated in a vacuum chamber. The warm seawater vapor is used as a very low-pressure working fluid to drive a turbine generator set to produce electricity. After passing through the turbine, the vapor enters a condenser. A direct contact or surface condenser permits cold seawater to condense the steam. If a surface condenser is used, the cold seawater flows on one side of a heat exchanger and the condensate forms on the other side. The condensate is desalinated water.

3.   HISTORY OF THE PROGRAM

Ocean Thermal Energy Conversion (OTEC) was one of six solar technologies in the original US solar energy program, which began under the sponsorship of the National Science Foundation in 1972. The Federal Ocean Energy Systems Program - as the Ocean Energy Technology Program was known until 1981 - was transferred to the Energy Research and Development Administration (ERDA) in 1975 and to the newly created Department of Energy (DOE) in 1977. Ocean energy research was part of a national strategy to reduce US dependence on oil imports.

As the OTEC R&D effort progressed during the mid 1970s, the Federal Program also began looking at whether ocean waves, ocean currents, or salinity gradients could be used to generate electricity. The Program placed most of this additional emphasis on a wave energy systems study, which yielded an experimental 125 kWe pneumatic wave energy conversion turbine by the early 1980s (Reference Guide to Fabrication, 1983).

Ocean energy research efforts led to other important accomplishments in the late 1970s and early 1980s. At-sea testing with Mini-OTEC (in 1979) (Burwell and Trimble, 1980) and OTEC-1 (in 1981) (Lorenz et al) demonstrated the technical feasibility of closed-cycle OTEC systems and provided data on heat exchanger thermohydraulics, environmental impacts, cold-water pipe deployment, and systems operations. These studies also indicated, however, that the technical and economic barriers to OTEC commercialization were more significant than anticipated.

OTEC research prior to fiscal year 1981 was directed toward large floating plant systems that would provide power to islands, certain parts of the mainland, and large plantships that produce energy-intensive products. Beginning in mid-FY81, interest in near- or on-shore systems increased. Research shifted from vertical cold water pipes to bottom mounted ones, and it shifted from floating systems to ones mounted on shore, on artificial islands or permanently fixed to near-shore shelves.

A design study of a 40 MW CC-OTEC plant was completed under joint sponsorship of DOE, industry and the State of Hawaii. A detailed preliminary plant design was completed (Ocean Thermal Corp., 1988).

Early in the federal OTEC research program, closed-cycle systems were believed to be closer to commercial readiness and received the bulk of attention. Open-cycle systems were thought to have significant long-term potential but were less developed and thus offered less potential for near-term production. In the 1980s, a government-wide shift toward longer-term, higher-risk research led to a redirection of attention from closed-cycle to open-cycle research (Federal Ocean Energy Tech. Program).

Presently the federal Ocean Energy Technology (OET) program focuses primarily on assessing and solving the technological problems leading to the development of land-based or near shore open-cycle OTEC systems ranging in size from 2-15 MWe. Researchers are looking at ways to make open-cycle OTEC systems work more efficiently - by designing smaller plants that produce more power, use smaller volumes of water and incorporate less expensive materials and components. Effort on surface contact heat exchangers has focused on material studies as they may apply to OC-OTEC and hybrid-cycle OTEC.

A number of recent discoveries (1983-88) have enhanced the economic prospects for both open- and closed-cycle OTEC systems. These include:

- Researchers have determined that plentiful and relatively inexpensive aluminium alloy can be used in lieu of the more expensive titanium for making heat exchangers to be used in the OTEC system (Qualification of Aluminium, 1988).

- Bio-fouling appears not to be a problem in cold seawater systems and can be controlled with minimal intermittent chlorination in warm seawater systems (70-ppb, 1hr/day) (Qualification of Aluminium, 1988).

- Scientists are developing cost-effective "state-of-the-art" turbines for open-cycle OTEC systems (Ongoing "Single Stage Steam Turbine Generator Design for an Advanced Open Cycle Development Project", 1988).

- In 1983, a test program of state-of-the-art industrial heat exchangers of 1 MWt rating was concluded at Argonne National Laboratory (ANL). Linde, Trane, Carnegie Mellon University, and Johns Hopkins University participated in the program. A wealth of information was obtained on the thermohydraulic performance of

various industrial enhancements: flutes, nucleation promoters,
wirewraps, and fins. It was demonstrated that increases in the
overall heat transfer rates of a factor of three over plain tubes
were possible.

- Researchers at the Sonar Energy Research Institute (SERI)
  developed a flash evaporator capable of converting warm seawater
  into low-pressure steam for use in an open-cycle OTEC steam
  turbine to produce electrical power. Evaporator efficiencies as
  high a 97% were achieved (Bharathan and Link, 1988).

- Desalinated water was produced by using the open-cycle process at
  the Seacoast Test Facility as a result of joint SERI/ANL efforts.
  This was the first time in the DOE/OTEC program that desalinated
  water was produced operating with proto-typical conditions (Thomas
  and Hillis, 1988).

- The deployment of a seawater supply system at the Seacoast Test
  Facility was completed. Co-operative fnding was provided by DOE,
  the State of Hawaii, and the Pacific International Center for High
  Technology Research (PICHTR). It provides the warm and cold
  seawater supply capability to allow research on near-shore open-
  cycle OTEC systems to evaluate electrical energy production at the
  165 kWe level and to evaluate cold water mariculture options
  (Lewis et al, 1988).

- Significant advances in direct-contact condenser application to
  OC-OTEC have been achieved. Packings have been experimentally
  tested in fresh water which provides thermal effectiveness in
  excess of 95% over the range of OTEC operating conditions.
  Computer models have been developed of the process. Agreement
  between predictions and experiments are excellent (Bharathan et
  al, 1988).

4.   MARKET ISSUES

4.1 Domestic

No commercial OTEC systems have been installed in the US. At present,
with adequate supplies of electrical energy and the price of oil
falling in most areas of the US, the near-term potential domestic
market is not large. If the OTEC industry is to succeed in selling
systems, there are several markets to be considered. Utilities in
Hawaii and the US island territories are most likely to use the first
systems. Here OTEC resources are plentiful and conventional energy
costs high. The second potential market is utilities in the US Gulf
Coast region. Here OTEC resources are available, but are not abundant.
Transmission costs will be high and OTEC will have to compete directly
with relatively cheap conventional fuel. Military installations in the
Caribbean and Pacific island offer another market for both electrical
energy production as well as utilization of the fresh water production

potential of OC-OTEC. The renewable, non-polluting aspects of OTEC
hold significant potential attraction for these applications. One
other market area in the islands which may have significant interest in
OC-OTEC are the hotel/resort developments. When cost analyses are
performed factoring in not only fresh water production but also space
cooling via utilization of the cold seawater into the electrical energy
production costs, the overall system economics improve dramatically
(Thermoeconomic Optimization of OC-OTEC Electricity and Water
Production Plants, 1985). Air conditioning using the cold seawater is
currently being implemented at the Natural Energy Laboratory of Hawaii
(NELH).

4.2   International

Viable OTEC locations, having a 20°C or more temperature differential
between surface water and water from the depths, can be found in many
areas of the world, particularly in the latitudes $\pm24^\circ$ from the
equator. Over 90 countries and territories are within 320 km of OTEC
resources. Many of them are energy importing nations and many are too
small for a commerical-sized nuclear power plant. For those countries,
OTEC may provide an attractive energy alternative (National Oceanic and
Atmospheric Administration, 1981).
      A study by John Hopkins Applied Physics Laboratory estimated that
the potential market in industrialized nations other than the US is
245,000 MW. The potential market in developing countries is an
additional 206,000 MW (Department of Energy, 1984).
      Currently Japan, and to a lesser degree France, continue to have
an active interest in the OTEC technology. Japan, in particular, is
aggressively pursuing the development of CC-OTEC for island market
applications. In the longer range, they have interest in OC-OTEC.

4.3   The Industry

Since no vigorous market exists for OTEC, no significant industry has
yet developed. To date, the ocean thermal energy conversion industry
has been predominantly composed of component manufacturers and
engineering firms, rather than companies which devote themselves
exclusively to OTEC. While a large number of companies may have an
interest in the future of OTEC, only a relative few are deeply
involved. There are, however, exceptions. Sea Solar Power and Marine
Development Associates, Inc. are examples of companies which were
created to commercialize OTEC and view it as their primary goal. The
recent turbine development solicitation (1988) by the DOE program to
develop an OC-OTEC turbine received inquiries from over 40
manufacturers and engineering firms. Current active involvement in the
program is provided by Mechanical Technologies, Inc. and Mitsubishi
Heavy Industries.
      Other companies which have made significant contributions to the
field include several engineering firms: Global Marine Development,
R.J. Brown, Gianatti and Associates, Harding Lawson Associates, E.K.
Noda and Associates, Makai Ocean Engineering, R&D Associates, and Gibb
and Cox, Inc. Several companies with experience in shipbuilding,

offshore oil, and similar industries have been involved - Hawaiian
Dredging and Construction, Alfred T. Yee, and Chicago Bridge and Iron
are some examples.

Large companies previously involved in the program were TRW Inc.,
Lockheed, General Electric, and Westinghouse.

4.4 Economic Viability

Since OTEC plants in the 2-15 MWe range appear to be the ideal size to
enter tropical island utility markets, private interest in this
technology should increase as technical problems are solved. Whether
industry actually undertakes the commercialization of OTEC, however,
depends on whether energy costs for this technology can compete with
those of conventional systems. This, in turn, depends on energy
economics, technical advances, and geographic location.

Recent cost of energy studies (Valenzuela et al, 1988) have
projected that the Federal OTEC system interim cost goals
(Table 1)(Federal Ocean Energy Tech. Program) can be achieved without
consideration of credits for production of fresh water or space
cooling. The relative subsystem cost breakdown for a 10 MWe plant is
shown in figure 1.

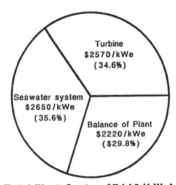

Total Plant Cost = $7440/kWe*

*1985 dollar   Ref: SERI/TR-253-3077 9/87

Figure 1. Cost breakdown for 10 MWe baseline OTEC plant.

4.5 Environmental Impact

Generally, OTEC technology is believed to have minimal environmental
impact in an absolute sense, i.e. its direct impact on the local
environment, as well as in a relative sense, i.e. by comparison to
other energy production technologies such as coal, oil and nuclear.
However, little data exists to support this claim. Research is
underway to obtain scientific data on the potential environmental
impact of several of the processes associated with the OTEC technology.
The key areas of uncertainty associated with potential environmental
impact are:

1.   The disposition of large quantities of warm and cold seawater
      required for the OTEC process; and

2.   The character and quantity of dissolved gases in the seawater
      which are released during the process.

TABLE 1. OTEC System Cost Goals (1985$)

	Present Technology	Interim Goal*	Long-Term Goal**
Total System Capital Cost (1985$/kWe)	10,000	7,440	3,300
30-Year Levelized Electricity Cost (1985$/kWh)	19	11	6

*   Competitive with oil-generated electricity on islands in 5
    years.

** Competitive with coal-generated electricity on mainland in 2010

      Potential environmental consequences of the seawater disposition
are all directly or indirectly related to induced changes in the
quality of the receiving waters. The OTEC discharges will be higher in
salinity and nutrients and lower in dissolved oxygen than the
groundwater or seawater into which they will be discharged. The
reduced oxygen content of the OTEC discharges may depress metabolic
activity in the coastal community. Nutrient-rich discharges may
stimulate the growth of macroalgae in the surge zone, which could be
aesthetically unpleasing to some observers in the immediate vicinity of
the discharge area. However, nutrient enrichment of the offshore
waters could be beneficial to the local fishing industry (Quinby-Hunt
et al, 1986).
      The chemistry of warm and cold seawater as regards nutrients and
dissolved gases is quite different, e.g. while warm seawater has
significant amounts of O_2, the cold seawater has much less; the pH of
warm seawater is approximately 8.2 at 25.5°C while cold seawater is
approximately 7.6 at 7.9°C. Salinity at the surface is significantly
greater than at depths. Major constituent dissolved gases in seawater
are O_2, N_2, and CO_2. Definitive measurements have been made to
determine the actual dissolved gas composition and nutrients (nitrates,
phosphates, etc.) (OTEC Environmental Benchmark Survey Off Keahole

Point, Hawaii, 1980). Since outgasing has an effect not only on the
environment but also on the system performance, investigations are
underway to measure these gases as well as to provide for their safe
disposal via reinjection to the seawater to be returned to the sea.
Measurements of O_2 have been made. Measurements of N_2 and CO_2 are in
the process of being made.

While the above potential environmental impacts cannot be casually
dismissed, it is believed that engineering will provide cost-effective
solutions.

5.   RECENT ADVANCES IN OTEC - TECHNOLOGY ISSUES

5.1 The OTEC System

The OTEC systems undergoing research have particular emphasis on Open-
Cycle OTEC. The key elements in an Open-Cycle OTEC power system
include the evaporator, the turbine, the condenser and an acillary
system, which in part provides for: deaeration, discharge of warm and
cold seawater, vacuum maintenance and exhaust. Analytical and
experimental work is currently underway for each of these subsystems.
Figure 2 shows a conceptual layout of a planned Net Power-Producing
Experiment (NPPE) having a nominal gross power of 165 kWe.

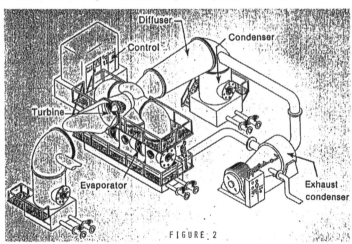

Figure 2.   Conceptual layout of a 165 kW experiment.

Investigations of the heat and mass transfer characteristics of
the evaporator and condenser system elements have been underway since
the inception of the program. Understanding of the phenomenon involved
has been advanced significantly and models have been developed.
Experimental correlations between analytical predictions and fresh
water experimental data have been good (Bharathan and Link, 1988;
Bharathan et al, 1988). A key issue being addressed by the current
experimental effort is that of the behavior of these subsystems in
seawater.

5.2 HMTSTA Test Facility

The Heat and Mass Transfer Scoping Test Apparatus (HMTSTA) was built to obtain data necessary for the design of a net power-producing system experiment (NPPE). The HMTSTA was engineered, fabricated, and installed by ANL at the Keahole, Hawaii site of the Natural Energy Laboratory of Hawaii, and is being used by SERI, ANL, and PICHTR to conduct open-cycle OTEC experiments with seawater. It consists of a barometric-leg spout evaporator, a shell-and-plate surface condenser, a dynamic vacuum system to maintain subatmospheric pressure, and a computerized data acquisition system (Fig. 3). The pressure difference between the evaporator and condenser is maintained by a throttling valve, which takes the place of a turbine-generator in an actual OTEC plant. The facility operates with approximately 100 l/s warm water and 63 l/s cold water. The overall design specifications are given in Table 2.

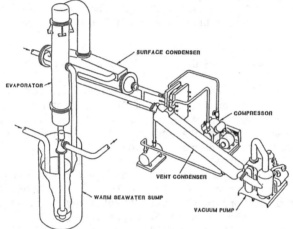

Figure 3. HMTST apparatus (Design by ANL).

TABLE 2. STF/HMTSTA overall design specifications

Warm Water Flow Rate	103 kg/s	(1600 GPM)
Cold Water Flow Rate	65 kg/s	(1000 GPM)
Rate of Steam Generation	0.5 kg/s	(3737 LB/HR)
Vacuum Pump Capacity	0.5 m³/s	(1000 CFM)
Evaporator Pressure	20 Torr	(0.4 PSIA)
Condenser Pressure	13 Torr	(0.25 PSIA)
Vacuum Pump Suction Pressure	11 Torr	(0.22 PSIA)

Phase 1 of the experimental program focused on the spout evaporator and surface condenser. Single- and multi-spout evaporator configurations were tested for a range of water loadings, flash pressures, and spout diameters. Evaporator effectiveness, multi-spout stability, vapor-flow resistance in the droplet zone, and droplet entrainment were measured. The surface-condenser performance was obtained by the Wilson-plot method to separate the vapor-side heat-transfer coefficient from the overall heat-transfer coefficient for a range of operating conditions. Mist elimination and deaeration tests were also conducted.

Phase 2, currently underway, focuses on direct-contact condensation, and a barometric-leg direct-contact condenser (DCC). The structure has been designed and built by ANL. The DCC internals have been designed and built by SERI. The DCC has been installed at the HMTSTA and is being used to obtain performance data for selected column packings under typical open-cycle OTEC operating conditions. Results from these and the earlier experiments will be used to refine and validate performance models that are essential for the design of the NPPE.

5.3 Preliminary Seawater Results from HMTSTA Evaporators

The evaporator thermal performance is a complex function of the spout and vessel geometry and water and steam flow parameters. The geometric parameters are the spout diameter, spout height, and interspout spacing, in the case of multiple spouts. The flow parameters include the spout velocity and the degree of superheat.

The evaporator results describe the thermal performance as a function of evaporator superheat (or temperature driving potential), the spout height, and the overall liquid loading. A measure of the evaporator thermal performance is the effectiveness, E, defined as the ratio as the ratio of the temperature difference actually achieved divided by the thermodynamically available temperature difference. The experimental results indicate that, in general, the thermal effectiveness increases with increasing superheat, height and decreasing liquid loadings. Figure 4 shows spout thermal effectiveness versus liquid loading for several spout diameters and for one multiple spout configuration.

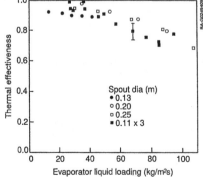

Figure 4.   Seawater spout evaporator. Influence of liquid loading.

The trend data obtained for the thermal performance for the simple spout using seawater are found to be similar to that obtained using freshwater in earlier tests completed at SERI. However, significant differences were observed in the behaviour of the steam/vapor above the seawater jet. Escaping steam and noncondensable gases in seawater tended to form a large layer of foam above the spout jet. Investigations are underway to ascertain the extent of the influence of foaming on the thermal effectiveness of the flash evaporation process for OTEC.

5.4 Preliminary Seawater Results from HMTSTA Surface Condenser

Testing of the HMTSTA surface condenser began in April of 1988. The surface-condenser system consists of a Stage 1 condenser (Fig. 5) where the majority of the steam is condensed, and a Stage 2 condenser (Fig. 6) designed to operate with a high ratio of noncondensables to steam. The Stage 2 condenser enhanced the thermal efficiency of the cycle and reduces the parasitic power required for the vacuum pump. Test parameters that were varied were water velocity, steam loading, fraction of steam condensed in the Stage 1 condenser, and outlet-steam saturation temperature in the Stage 2 condenser. For each series of tests, the overall performance was determined in terms of the heat-transfer coefficient and steam-side pressure drop.

The following observations can be made on the basis of a preliminary analysis of the test results:

1. the thermal performance of the Stage 1 condenser is within the predicted range;

2. no significant maldistribution of the steam/gas mixture was observed;

3. the thermal performance of the Stage 2 (vent) condenser was lower than predicted, indicating that further experimental verification may be required; and

4. the steam-side pressure drop is small and predictable.

Figure 5. Rosenblad surface condenser (Courtesy of ANL)

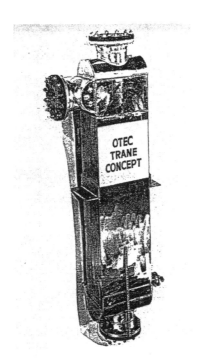

Figure 6. The Trane heat exchanger (Courtesy of ANL).

5.5 Corrosion and Bio-fouling - Interim Results

Research in bio-fouling and corrosion was carried out at NELH by ANL
during the period 1981-1986. This laboratory was protected from wave
and surf action and supplied with warm and cold seawater of open-ocean
quality. The initial test emphasis was on tubular materials (titanium
and alloy steels) known for good resistance to seawater corrosion and
suitable for abrasive cleaning techniques (brushing and recirculating
sponge balls). As the tests progressed, the emphasis switched to
aluminium. The major findings of the ANL test program were:

1.   fouling from deep cold seawater is negligible (Fig. 7);

2.   intermittent chlorination at environmentally acceptable levels
     prevents the buildup of bio-fouling film and removes existing film
     from both plain and moderately enhanced surfaces (Fig. 8); and

3.   selected alloys of aluminium (e.g. 3003 and 5052) can be predicted
     with 90% confidence to have long service lives (15+ years) in
     seawater. (Figs. 9 and 10). These findings opened the door for
     non-traditional OTEC materials and geometries to be considered for
     OTEC service.

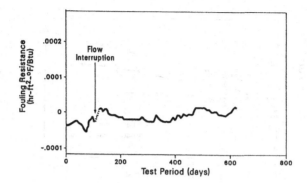

Figure 7.   Free fouling of stainless steel test section in cold water.
            (Data provided by ANL)

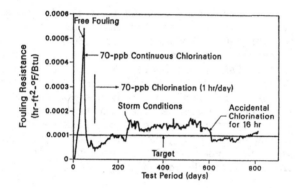

Figure 8.   Bio-fouling management is similar for spirally fluted and
smooth tubes.   (Data provided by ANL)

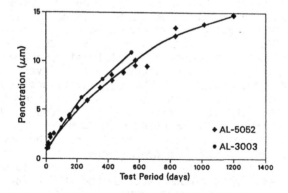

Figure 9.   Aluminium corrosion rate in cold water.
            (Data provided by ANL)

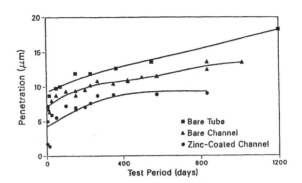

Figure 10. Warm-water aluminium-3003 corrosion is compatible with long component life. (Data provided by ANL)

Tests at ANL of heat-exchanger designs considered promising for OTEC has previously shown that brazed-aluminium "compact" exchangers not only demonstrated higher performance under OTEC conditions, but were also much less expensive than shell-and-tube units made of traditional corrosion-resistant materials. When the bio-fouling studies confirmed that irregular configurations could be considered for OTEC heat exchangers, and the corrosion studies demonstrated that aluminium is suitable for seawater service, interest switched to this material. A net result of this research in bio-fouling and corrosion is that the cost of the heat exchangers in a closed cycle OTEC plant is now estimated to be of the order of 30% of the total plant cost, as compared to an estimated 50% a decade ago.

5.6 Production of Desalinated Water Using OTEC Processes

OTEC processes that employ the flash-evaporation of seawater are capable of producing fresh water as a by-product of power production. In many locations, fresh water has a market value in the range of $1-11 per 3875 liters (1000 gallons) (Thermoeconomic Optimization of OC-OTEC Electricity and Water Production Plants, 1985). The price for OTEC electrical power becomes competitive with fossil-produced power when water is also produced and sold at these prices. Even when electrical power is not desired, open-cycle OTEC appears attractive for the production of water alone. It has been estimated (Thermoeconomic Optimization of OC-OTEC Electricity and Water Production Plants, 1985) that the cost of fresh water produced to OTEC may be one-third that of fresh water produced by the standard multi-stage flash-evaporation process using fossil energy as a heat source.

The technical feasibility of using ocean energy to desalinate seawater was dramatically demonstrated by the HMTSTA in August of 1987. This apparatus, located at the STF/NELH, produced 22 l/s of potable water having a salinity of 86 ppm, one-fifth that of local tap water available at the test site. It was the first ever conversion of seawater to fresh water using the thermal energy stored in the seawater itself.

5.7   OTEC Turbine Developments

The turbine for the open-cycle OTEC represents an area of technical uncertainty and one of the more costly components. Estimates of turbine cost for a 10 MW power system amount to 35% of the total plant cost of around $5600/kW (Thermoeconomic Optimization of OC-OTEC Electricity and Water Production Plants, 1985). The goal of the DOE Multi Year Program Plan (MYPP) is to reduce this cost to $3200/kW (Federal Ocean Energy Tech. Program).
       Because of the small expansion head available to an OC-OTEC turbine, the power output per unit actuator area is low, resulting in a large turbine. For the same reason, the structural and dynamic requirements of the rotor and blading design may be uniquely different from existing hardware. further, the physical and chemical composition of the steam in terms of droplet and chloride content may depart significantly from utility practice and material compatibility issues may arise. Fluctuations in the seawater flow rates and temperatures also may create control and stability concerns in the operation of the turbine.
       Several design studies have been completed in the program which indicate strongly that existing technology and, possibly existing hardware, can meet the aero- and thermodynamic and structural requirements for an OC-OTEC turbine up to approximately 2 to 5 MWe. This conclusion was confirmed by a panel of turbine industry experts which reviewed the technical requirements of the OTEC application and helped to develop a specification for the procurement of the turbine test article for the NPPE (Ongoing "Single Stage Steam Turbine Generator Design for an Advanced Open Cycle Development Project", 1988).
       The turbine development is being implemented by a dual path approach. One path deals with the development of the specific turbine configuration to be used in the NPPE. This activity involves two efforts:

(a) the development of a turbine using existing engineering
     technology; and

(b) the assembly of a turbine by limited modification of existing
     turbine hardware/systems.

The second path is study of innovative approaches to turbine requirements for large turbines to be used in multi-megawatt systems.

5.8   Seawater Supply Systems (SWSS) for OTEC

As noted earlier, seawater supply systems comprise a major cost element of the OTEC system (Fig. 1). Historically this subsystem has been a major concern and has been analytically studied as well as being the focus of several experimental efforts involving actual pipe deployments. The most recent program efforts in this area have been minimal because

(a) funding restrictions mandate a prioritized approach to addressing the technical issues; and

(b) the required seawater supply system for the NPPE exists at the DOE/Seacoast Test Facility (STF).

There are several SWSS deployed at the NELH. Two of these systems will be utilized by the DOE/STF. One system supplies up to 100 l/s of warm seawater and 63 l/s of cold seawater to the HMTSTA. A second system will supply up to 600 l/s of warm seawater and 410 l/s of cold seawater to the NPPE.

The deployment of these SWSS have added significantly to engineering practice in this area. There are five polyethylene pipelines, ranging from 30 cm to 100 cm, capable of drawing water from approximately 700 m depth off Keahole Point, Hawaii. Various deployment techniques were used. In one deployment scheme the pipe was assembled on shore as a continuous single length pipe. The pipe was then pulled over a series of rollers into the ocean along a prescribed path by a tug. Weights were added on shore along the length of the pipe to facilitate sinking and anchoring. In another scheme, the deployment operations consisted of the following major steps: assembly on shore in 3 sections; each section floated air-filled on the surface with one of the three sections having distributed anchor weights; two sections were joined into one and then the resulting 2 sections towed as a single line over the predetermined pipeline route; and final joining was made of the remaining section and controlled flooding/lowering to the bottom position down to 700 m. The key to the success of this operation has been the careful detailed planning and controlled submergence of large masses over a relative short time (approx. 3-5 days). (Van Ryzin et al, 1988).

In anticipation of issues which will need to be addressed for multi-megawatt OTEC SWSS, a workshop conducted by PICHTR provided expert critique of the fabrication and deployment options for the seawater systems of land-based OTEC plants. Conduits in a diameter range of 3-6 m were considered, and means to achieve significant cost reductions were discussed. Design concepts and deployment techniques available were examined. It was concluded that:

1.   the design and deployment of extruded polyethylene pipes less than 1.6 m in diameter is state-of-the-art; and

2.   the fabricability and cost effectiveness of larger pipes has not been demonstrated.

It was determined that deployment and in-place bending stresses must be more accurately predicted. Furthermore, very different deployment schemes are envisioned for OTEC if large diameter single-length or segmented pipes are considered. In the one case, deployment

stresses will be critical and, in the second, mating and joint
attachment issues will predominate.
    Three groups were formed to focus on the following areas:

1.   near-shore zone;

2.   conventional materials and offshore techniques; and

3.   new materials and concepts.

The recommendations from this effort will be published (Nihous and
Vega, 1988) shortly and will also form the basis for developing a long
range R&D program.

5.9   Plans and Activities for the Future

Testing will continue on the existing HMTSTA configuration (Fig. 3) to
complete the evaporator, surface condenser, and deaeration
investigations.
    In a second phase of testing on the HMTSTA, a direct contact
condenser (DCC) will be utilized in place of the surface condenser. In
the direct contact condenser, the steam will be condensed by direct
contact with cold seawater. Various packing configurations will be
investigated. The data collected in the HMTSTA will be utilized to
help design an experiment which will ultimately lead to the NPPE. The
NPPE will utilize up to 600 l/s of warm seawater and 410 l/s of cold
seawater, and will produce approximately 165 kWe of gross power. The
objectives of this OC-OTEC system experiment are to:

1.   obtain data on system performance;

2.   obtain system control data and experience;

3.   evaluate system stability;

4.   obtain subsystem performance data; and

5.   initiate the addressing of scaling issues.

Major program activities and their time frames are shown in figure 11.

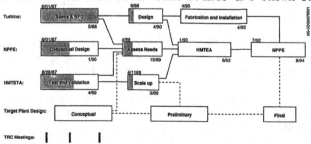

Figure 11.   Research thrusts: net power producing experiment (165 kWe).

6.   REFERENCES

Bharathan, D. and Link, H (1988) 'Preliminary results of seawater
    evaporation in single and multiple spouts`, SERI Letter Report.

Bharathan, D. Parsons, B.K. and Althof, J.A. (1988) 'Direct-
    contact condensers (DCC) for open-cycle OTEC applications`,
    SERI/TR-252-3108.

Burwell, D.N. and Trimble, L.C. (1980) 'Mini-OTEC: An operational
    OTEC plan`, National Conference on renewable energy
    technologies, Honolulu, HI (available from NTIS, PC A99/MF A01).

Department of Energy (1984) 'Comprehensive ocean thermal
    technology application and market development (TAMD) plan`,
    First Annual Update, p.4,9.

Lewis, L.F., Van Ryzin, J. and Vega, L. (1988) 'Steep slope
    seawater supply pipeline`, Proc. American Society of Civil
    Engineers 21st Int. Conf. on Coastal Engineering, Spain.

Lorenz, J.J. et al 'OTEC-1 power system test program: Performance
    of one-megawatt heat exchangers`, Ocean Thermal Energy
    Conservation Program, Argonne National Laboratory, ANL/OTEC-PS-
    10.

National Oceanic and Atmospheric Administration (NOAA) (1981)
    'Ocean thermal energy conversion`, Report to Congress Fiscal
    Year 1981, p.32.

Nihous, G.C. and Vega, L.A. (1988) 'Bottom-mounted OTEC seawater
    systems workshop summary report`, PICHTR, (Available from the US
    DOE).

Noda, E., Beinfang, P.K. and Ziemann, D.A. (1980) 'OTEC
    environmental benchmark survey off Keahole Point, Hawaii`, Univ.
    of Hawaii.

Quinby-Hunt, M., Wyle, P. and Dougler, A. (1986) 'Potential
    environmental impact of OC-OTEC`, Journal on Environmental
    Impact Assessment Review, Vol. 6, pp.77-93.

Thomas, T. and Hillis, D. L. (1988) 'First production of potable
    water by OTEC and its potential applications`, Argonne National
    Laboratory, Oceans '88, Honolulu, Hawaii.

Valenzuela, J.A., Jasinski, T., Stacey, W.D., Patel, B.R. and
    Dolan, F.Y. (1988) 'Design and cost study of critical OC-OTEC
    plant components`, SERI/STR-252-3246.

Van Ryzin, J.C. et al (1988) 'The HOST-STF (OTEC) project in
    Hawaii: Planning, Design, Construction', Proc. PACON, Honolulu,
    Hawaii.

'A Reference Guide to Fabrication Details and Preliminary
    Shakedown Testing of a Pneumatic Wave Energy Converter'
    (1983) SERI/TR-252-1765 DE84000004.

'Federal Ocean Energy Technology Program. Multi-year program plan
    FY85-89' US Department of Energy; Wind/Ocean Technologies
    Division; Solar Electric Technologies; Assistant Secretary,
    Conservation and Renewable Energy; DOE/CH/10093-100.

'Qualification of aluminium for OTEC heat exchangers' (1988)
    Internal Report, Argonne National Laboratory.

'Single Stage Steam Turbine Generator Design for an Advanced Open-
    Cycle Development Project' (1988), Ongoing Subcontract No.
    18072-01.

'Thermoeconomic optimization of OC-OTEC electricity and water
    production plants' (1985) SERI/STR-251-2603 DE85121299.

'40 MWe OTEC power plant' (1988) Ocean Thermal Corporation.

PART V

Living Resources

LIVING MARINE RESOURCES, TECHNOLOGY AND THE EXTENDED ECONOMIC ZONE

JOHN E. BARDACH
Environment and Policy Institute
East-West Centre,
Honolulu, Hawaii,
U.S.A.

1. INTRODUCTION

The net and the hook are among man´s oldest inventions. With some
related tools, like tongs, rakes and spears also dating from the dawn
of history, they are still the implements with which man renders into
his possession what edible, and otherwise useful, bounties the sea has
to offer. The hooks were once made of bone, stone, or wood, now they
are made of steel, and the nets were made of plant fibres while they
are now fashioned of often all too permanent plastic materials.
Engineering acuity has not changed the basic concepts of these
implements except for making them more durable, and, in the case of
nets, larger, deeper and longer. The enormous technical advances in
fishing, over the ages though, lie in the ships that take the fishermen
to sea, the means by which fish are found, the mode in which fishing
tools are deployed, and the ways in which the harvest of the sea is
preserved and distributed. These latter aspects of fishing have
influenced fishing to such an extent that they are in essence
responsible for the extension of man´s domain out to 200 nautical miles
from shore.

Whatever the reasons for agreeing on 200 nautical miles as the
extent of the EEZ that limit certainly more than bounds the locations
in the world´s seas where virtually all life can be found (Fig. 1).
Only about 5% of sea life in accessible surface waters occurs in the
open ocean, outside EEZ boundaries; in fact, most fish and
invertebrates occur much closer to shore. They are unevenly
distributed under the influence of various geographic and climatic
factors (Atlas of the Living Resources of the Sea, 1981) as will be
dealt with later, but they are in the aggregate of great importance to
the well-being of mankind, supplying it with about a quarter of animal
protein intake.

The yield of fisheries has increased, in part due to technical
improvements that come from making war at sea, and while fish finding
(for some species) can now be done with sonar and while very large nets
and lines of many miles can be deployed, the common property nature of
the resources impose on us the need to attend to the social and
institutional parameters of fishing, so that not more stocks become

D. A. Ardus and M. A. Champ (eds.), Ocean Resources, Vol. I, 235–249.
© 1990 *Kluwer Academic Publishers. Printed in the Netherlands.*

overharvested than have been in the fairly recent past, and renewable
resources remain for sustained future use.

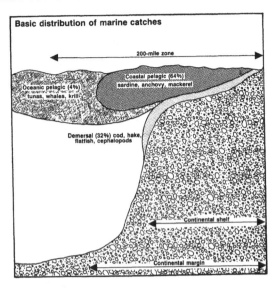

Figure 1. Most accessible Marine Living Resources are well inside the
200 mile EEZ.

Try as one may, though, wise management of territorial seas and
EEZ fishery resources will not satisfy the demands of a much larger,
and in the aggregate, probably more affluent world population, thus
aquaculture as opposed to capture fisheries will gain greatly in
importance. It will include techniques for the reproduction feeding,
containment, and sanitation of many different marine organisms, and
pose challenges for the fashioning of various new technologies, now
barely begun to be deployed in the marine realm. After touching upon
the distribution of renewable marine resources and upon selected
aspects of fish finding, fish catching and fish saving, this article
will end with a look at likely aquaculture development and some of its
technical exigencies.

2.   WHERE THE FISH ARE, HOW THEY ARE FOUND AND HOW THEY ARE CAUGHT

Large portions of the globe, seen from on high, are blue, which has
sometimes been called the desert colour of the sea. Except for a rare
passing school of tuna, and for here or there a coral island that
conserves its nutrients in its shallow brightly lit waters, the very
open, blue ocean is indeed not rich in harvestable life. In fact, the
open tropical surface waters, which constitute nearly 50% of the seas'
total extent, are permanently sealed from more fertile deeper waters by
a thermal discontinuity (thermocline). The world's fertile fishing
regions are found in the shallow epicontinental seas (e.g. North Sea,
East China Sea) and in places that receive nutrients from large rivers
(e.g. Gulf of Mexico, Bohai Bay). Advection of nutrients, necessary

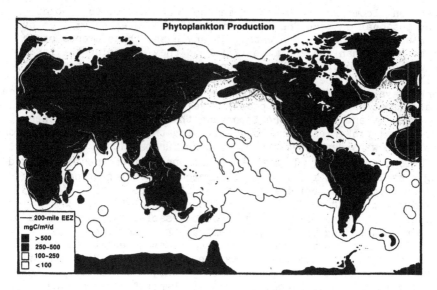

Figure 2. Phytoplankton production in the sea determines where Living Resources are.

How the world's two marine fishing industries compare

	Large scale	Small scale
Number of fishermen employed	Around 500,000	Over 12,000,000
Annual catch of marine fish for human consumption	Around 29 million tonnes	Around 24 million tonnes
Capital cost of each job on fishing vessels	$30,000–$300,000	$250–$2,500
Annual catch of marine fish for industrial reduction to meal and oil, etc.	Around 22 million tonnes	Almost none
Annual fuel oil consumption	14–19 million tonnes	1–2.5 million tonnes
Fish caught per tonne of fuel consumed	2–5 tonnes	10–20 tonnes
Fishermen employed for each $1 million invested in fishing vessels	5–30	500–4,000
Fish destroyed at sea each year as by-catch in shrimp fisheries	6–16 million tonnes	None

Figure 3. These are two fishing industries in the world; large and small scale or industrial and artisanal.

for plankton blooms as the basis of marine food chains, is brought
about by seasonal or more frequent mixing of the water column, or by
nutrient upwelling due to currents. In any case, these events occur
closer to shore than the 200 mile limit and they determine the location
of the world's important fishing grounds (Fig. 2). Inasmuch as the
major landmasses of the globe, and with them most semi-enclosed seas
and large bays, as well as major rivers, lie in the Northern Hemisphere
where also most of the world's people live, fisheries north of the
equator are, by and large, more important than those in the southern
oceans. Thus, fisheries technology following the industrial revolution
and the world wars developed out of western Europe, North America, the
Soviet Union, and Japan. Today, these nations register most motorized
fishing vessels, while millions of fishermen in developing nations
still fish from small, primitive craft (ICLARM Report, 1987) (Fig. 3),
the balance begins to shift. A mix of foreign assistance,
streamlined techniques of building plastic vessels, outboard motors of
varied horsepower, and the like, as well as the removal of much
competition due to 200 mile EEZs, begin to alter somewhat the balance
of sites of fishery production. This trend is shown in FAO statistics
where what one might call "old fisheries" - they are all too frequency
over exploited - have declined, and "new fisheries", some of them in
the tropics, have shown spectacular increases (Fig. 4) (FAO Yearbook,
1988).

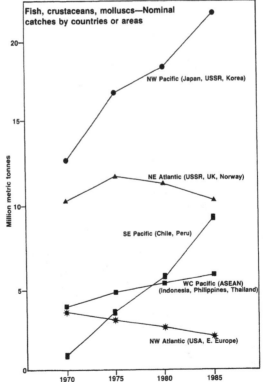

Figure 4. Contrast between "new" and "old" fisheries in the world; the
former are declining, the latter are on the rise.

The increases have generally been due to the tapping of less exploited stocks and use of better technologies. Examining these and starting at the lower end of the scale, near the artisanal end, one might well put first improvements in energy efficiency of motors and engines of all kinds and attention to new antifouling materials and paints. One would also look to the use of motorsailers, perhaps of a catamaran type, perhaps to cheaper radio-direction finders, and generally an evermore pervading use of plastic materials. Leaving aside shore installations, which can often stand improvement, especially in the provision of ice for the catch, portable and well foldable cranes for unloading containers, and improvements in the power winches for the handling of nets, find their way onto smaller vessels also. There is a need for better filleting machines, and where applicable for reason of the type of catch, for devices and procedures to begin on board the stabilization of the bycatch (i.e. that portion of the catches which are generally thrown overboard because they do not bring a good price).

These innovations may sound fairly prosaic, but they would, in the aggregate, make fishing more cost effective, as well as safer, and contribute to conservation. For the finding of fish, it is still the experience and the skill of fishermen, paired with inputs from ocean science, which will count the most, even though sonar can best help under certain conditions, that is, when fish schools are dense and in mid-water. Being based on transcription into a trace of the density, discontinuities between the swimbladders of fish schools and the water, its effectiveness is limited by the definition that is attainable under economically acceptable conditions. That definition has steadily improved, and as sonar devices can now scan horizontally (sidescanning) as well as vertically, one may certainly look in fisheries sonar devices to the same kind of evolution - in capability, size and price - that has characterized other electronic instruments. Another approach to fish spotting may come to lie in the use of automatically transmitting chemical sensors deployed from buoys. The idea is derived from the way sharks and tuna sense their schooling prey, the slime of which contains chemicals that dissolve or disperse in the water. If such buoys could be economically manufactured, and a large number dispersed over a good portion of likely fishing grounds, the time spent in search could be greatly cut.

Nets, as such, are already very effective, in some cases, even too effective, they are likely to change least through new technologies, perhaps because they have been around so long that most basic innovations on them have already been made. Fishing vessels, however, vary greatly in size and in design, depending on where they go and what they hunt. In coastal fisheries, there seems to be a trend, though, towards vessel design that can serve more than one purpose and, together with greater engine efficiency and use of plastics, a tendency towards more broad-beamed, sturdier vessels than were built some decades ago. Tuna seiners, longliners, and trawlers for bottom or near bottom fish, such as hake and pollock though, remain relatively specialised.

The trends in trawlers and fishing vessels that can be ascertained from the Lloyds Register are,

(1)   A consistent increase in the gross tonnage (g.t.) and numbers of
      fishing vessels in the past seventeen years,

(2)   Most increases occurred in the 1000-2000 g.t., and in the above
      4600 g.t. ranges while the smallest listed vessels of 100-500 tons
      showed less spectacular increases,

(3)   The first and second places were consistently occupied by the USSR
      and Japan respectively, while the USA moved from 8th to 3rd place
      between 1970 and 1987 (Table 1) (Lloyds Register).

These changes, especially point 2, illustrate that the world's
foremost fishing nations exploit their EEZ's more thoroughly even while
continuing their quests of distant water fishing. Some of the larger
vessels are superseiners for tuna (USA) while other even larger ones of
10-20,000 g.t., mostly under the flags of the USSR and Eastern European
nations, are trawlers and seiners with substantial processing
capabilities on board.

TABLE 1. Ranking of the Top Ten Fishing Nations by tonnage
and number of vessels over 100 gross tons*

	1987	1986	1985	1980	1975	1970
USSR	1	1	1	1	1	1
Japan	2	2	2	2	2	2
USA	3	3	3	4	4	8
Spain	4	4	4	3	3	3
South Korea	5	5	5	5	7	-
Norway	6	7	7	7	9	7
Poland	7	6	6	6	6	5
Denmark	8	-	-	-	-	-
Peru	9	8	9	-	-	-
Canada	10	9	-	-	-	10
Cuba	-	10	8	8	-	-
France	-	-	10	10	8	6
United Kingdom	-	-	-	9	5	4
West Germany	-	-	-	-	10	9

* Compiled from Lloyd's Register, by Hal Olson, EAPI

      In fact, as fish saving best begins with fish catching there is a
tendency to initiate processing on board and to more and more transfer,
at sea, to processing or refrigerated carrier vessels. Unitized
transferable containers are increasingly employed, as aré mechanical
devices, such as pumps to empty nets of their burgeoning contents. It
must be repeated, though, that the differences between mechanized and
artisanal fisheries are enormous, that they will most certainly
persist, and that for the greatest number of fishermen in the world,
technological advances - if they can afford them - are of a very
elementary kind, such as an ice supply on shore, plastic boxes to store
the catch, services to warn them of storms, engine parts for

replacement, and the like. More important than those improvements, though, will be attention to a reduction in the numbers of artisanal fishermen through culturally acceptable socio-economic measures to be invented by many nations. This kind of "social fishery technology", important as it is, is, however, outside the realm of this discussion.

3.   FISH SAVING

The use of ice to keep fish from going bad is an old practice, while curing, salting, pickling, drying and smoking the easily spoilable flesh of aquatic animals, is older still. Today only about 20 percent of the catch of food fish is marketed fresh, usually chilled, while 23 percent is frozen (FAO Yearbook, 1988). Food habits, among other factors, dictate that the fresh fish market predominates in much of Asia, while freezing is now a well established practice in Europe, North America and Japan. The rest of the catch for food is variously treated, from canning (salmon, tuna, sardines, 12.4%), to curing, to smoking and the like (14.7%). About a third of the total world catch (29.5%), consisting of small oily species at the base of the food chain, is rendered into fishmeal and oil and used in animal feeds and industry. Opinions differ whether or not a concerted effort should be made to invent techniques for using de-fatted mince of such fishes, as a base for fish sausage and the like, and to increase, thereby, the volume of ocean harvest that can be eaten directly, or whether the use of fish meal for additives to the feed of domesticated stock - from chicken to shrimp or salmon and trout - is a better use. Eventually, as demand for seafood rises, some attempts at such upgrading in the utilization of certain species will, no doubt, be made along patterns largely pioneered by Japanese surimi (a kneaded fish product) and kamaboko industries. The once low value Alaska pollock already experienced such use of partly chemical, partly mechanical, transformation of its flesh into fake crab or lobster claws. Its fishery is now the largest single species exploitation known; it will surely pose a challenge for management towards its sustainable yield. Before menhaden, anchovies, and similar plankton feeders become upgraded to be food specialities, fish saving technology is likely to turn its attention to better use of the great amount of fish that is now discarded, either at sea, or at the point of first dealing with the catch through filleting, a process where up to 50 percent of the weight may now be discarded. The discard or bycatch, predominantly of shrimp and other fisheries for high value species, may consist of many kinds of fish of various sizes, it is believed to amount, world-wide, to about five million metric tonnes, more than the total food fish catch of the United States.

Saving this large amount of protein from being, in essence, shark food and thus waste, poses difficult problems for food technology because of shapes, sizes, and differences in the chemical composition of the raw material. The methods devised for this upgrading of bycatch and offal will have to be cheap and simple, inasmuch as much of the practices that give rise to it are in the tropics. Some beginnings towards perfecting such practices have been made mainly in Asia, but as

the products are new and unaccustomed, the rate at which they gain
acceptance will depend on changing tastes and economics, as much as on
technical inventions. In view of the importance of fish in Asian diets
it may be of interest here to examine some training priorities listed
by the Southeast Asian Fisheries Development Centre (Table 2)
(Newsletter, SEAFDEC, 1988).

TABLE 2. Ten training subjects required in the field of
fishery post-harvest technology, listed by the priorities
indicated by the Member Countries in Southeast Asia.

Priority	Subject
1.	Fish Handling
2.	Packaging of Fish and Fish Products
3.	Preservation of Fish and Fish Products
4.	Analytical Methods for Fish and Fish Products
5.	Quality Assessment of Fishery Products
6.	Microbiological Techniques for Quality Assessments
7.	Fish Processing
8.	Food Engineering
9.	Post-harvest Handling of Live Aquatic Animals
10.	Product Development

While on the subject of fish food technology, krill exploitation
must be mentioned. Decades of krill research demonstrated that there
are indeed large tonnages of these small crustaceans in Antarctic
waters, but they vary in prevalence and in location from year to year.
Once seined, they are very easily bruised, prone to undergo rapid
autolysis, and discolouration. Their gut must be removed before
processing because the algae they contain tend to cause digestive upset
in mammals. In short, they are very difficult to process into
acceptable products apart from the fact that they are indeed at the end
of the earth, as it were, that they are only accessible for part of the
year, and that petroleum costs influence their exploitation more than
is the case with other fisheries. Apart from all this, unresolved
questions in the Antarctic Treaty, especially in regard to 200 mile
EEZs and the total conservation of Antarctic aquatic ecosystems, will
have a bearing on future krill exploitation.
        Looking into the near term, i.e. the twenty to thirty years future
of fish saving, fishery technology expert Paul Hansen of Denmark
confirms as follows some of the things said about it above (Advances in
Fish Science and Technology, 1981).

"The world fisheries of the future are likely to include more
species and a larger proportion of small fish and shellfish than
in the past. A number of small, and medium size fish, which up to
now have been reduced to fish meal and oil, will be preserved for

food in the future. This will require new and improved
preservation methods both at sea and ashore.

Recent advances made in the bulk handling of small fish indicate
that the chill preservation at sea may be greatly improved. More
vessels will be equipped with chilled or refrigerated sea water
systems, and some may supplement these systems with carbon dioxide
in order to extend the storage life of the catches. Systems for
ice packing of larger fish at sea are capable of improvement,
particularly in the tropical areas where efficient ice packing may
provide storage lives of several weeks. The fishing vessels of
the future will be equipped to grade their catches and stow them
in boxes, containers, and tanks according to particular
requirements.

The larger fish of well known species will tend to be reserved for
fresh fish markets or for high-price delicatessen products. A
number of such products will be developed and retail packed for
distribution via supermarket sales cabinets operating near 0
degrees C. Some of these products will combine two or more
preservation methods such as salting, pH adjustment and heat
pasteurization. Chemical preservatives, on the other hand, will
probably be restricted even more than today. The industries
producing cheap standard products will have to rely to an
increasing extent on smaller fish and shellfish of less attractive
species. Machinery for heading, gutting, filleting, skinning and
deboning will be developed for these species, and the edible parts
will be preserved by canning, curing or freezing. Canning will
remain very important for the preservation of the small fatty
fishes, which could also be salt cured much more than today
provided that larger markets are developed. It is expected that
mince of the less oily species will be dried on a large scale to
provide cheap high protein products."

4.   FISH AS FOOD AND AQUACULTURE

Proteins of aquatic provenance - mostly fish, but also crustaceans,
molluscs, and the like - make up about one-quarter of all animal
products for direct human consumption. Milk and its products and eggs
are included in these comparisons. The roughly 30 million metric tons
of fish (1986 figure) that are annually turned into fish meal and oil
primarily as ingredients for livestock feed and even for some feeds for
aquaculture are left out in this comparison. Asia, particularly
Southeast Asia and China, is by far the most important location for
aquaculture (see Tables 3 and 4). In terms of species groups, finfish,
primarily from freshwater locations, supply the largest amounts of
fisheries products with molluscs and seaweeds not far behind;
crustaceans bring up the rear but are gaining fast (see Table 5)
(Aquaculture - moving from craft to industry, 1988).
   Reasons for the preponderance of fish in freshwater aquaculture
are that fish are, by and large, more difficult to contain in the sea

than in ponds, that freshwater fish, or those that spawn there
(salmon), have larger eggs and are easier to rear, and that pond fish
culture dates back over millenia in China and Europe. In contrast most
molluscs, as well as seaweeds, occur in the sea and are sessile and
need only attachment, and not containment devices.

TABLE 3. Summary of Aquaculture Production by Food and Agriculture
Production by Food and Agriculture Organisation Regions (1983 data)

Regions	Finfish	Molluscs	Crustaceans	Seaweeds	Totals
Africa	43,553	286	26	-	49,865
Asia/Pacific	3,357,978	2,586,464	75,644	2,392,045	8,412,131
Europe/Near East	726,530	494,719	162	100	1,221,511
South America	167,797	30,883	20,188	1,637	220,505
North America	152,088	133,178	27,425	-	312,691
Totals	4,447,946	3,245,530	123,445	2,393,782	10,216,703

Source: C. Nash, "Aquaculture Attracts Increasing Share of Development
Aid", Fish Farming International, June 1987, 21-24.

TABLE 4. Importance of Fish and Aquaculture in Asian Nutrition

TABLE 5. Contributions for Aquaculture to the World Supply of Fisheries Products (1983 data)

Source	Finfish	Crustaceans	Molluscs	Seaweeds
		(metric tons)		
Capture fisheries (less aquaculture)	62,753,000	3,069,000	878,000	811,000
Aquaculture fisheries	4,448,000	123,000	3,246,000	2,394,000
Percent share of aquaculture fisheries in world supply	7	4	79	75

Source: C. Nash, "Aquaculture Attracts Increasing Share of Development Aid", Fish Farming International, June 1987, 21-24.

Some notable developments of aquaculture in the marine realm, or mariculture if you like, are rapid advances

(1) in the spawning of several marine fishes such as flatfishes, yellowfins, and mahi mahi (or dolphinfish as they are also called)

(2) the combination of fish shelters with stocking

(3) the rearing of giant clams and other tropical shallow water species that could be tended

(4) another is the use of deep, cold, clean, nutrient-rich ocean water for certain aquaculture practices.

The latter, grown out of experimentation with OTEC, has some especially interesting properties for aquaculture. The cold water taken from below the thermocline contains no fungi or bacteria or other fouling organisms, and the aquaculturist using it, has thus a leg up, as it were, on the problem of sanitation, and an option of raising cold water species such as salmon and abalone. Demand for the former may be a problem in the tropical sites where such culture is possible, especially in as much as many potential sites for salmon rearing are being developed in cold (also relatively clean) waters in Northern Europe and North America.

Abalone rearing depends, among other things on having algal food for these voracious herbivorous molluscs. That food can be produced well, it seems, in so-called tumble culture where absorption of the nutrients by the algae is a function of the rate at which they are bathed by large amounts of water, mimicking as it were, the action of

the tides to which these algae are subjected in the natural
environment. Warm water filter feeding molluscs could, of course, also
be grown.
        Aside from its aquaculture potential, the cold water is the sine
qua non of energy generation, can provide air conditioning and fresh
water. Multiple use systems are thus likely to be invented at future
OTEC sites.
        Most marine food fish species are carniverous and the cost of
feeding them can make up half or more of the total rearing costs.
Development of cheap feeds and techniques of administering them with
computer controls, already in use in Norwegian salmon culture, are
likely to become widespread in marine fish culture.
        Combinations of shelters which attract small forage fishes,
training to accept food on sensory clues (mostly sound), combined with
stocking, is practiced in Japan where cooperatives control certain
extended portions of the near shore under water terrain. The
ecological relations and economic advantages of this kind of site
management cum aquaculture are not yet well worked out, but it appears
that substantial improvements over the "wild state" are possible.
        The same is likely to be true on coral reefs where the rearing of
giant clams has already begun. These molluscs of the family
Tridacnidae are an intriguing case indeed. The clams grow slowly but
to a large size, and their mantle is edible, the flesh preferably dried
to withstand spoilage. Most of their flesh could become something of a
protein staple, while the shell-closing adductor muscle is among the
most prized seafoods. Wealthy Asian gourmets maintain it carries
aphrodisiac properties, and demand has led to the decimation of giant
clam beds by tuna fishermen who poach over a part of the clams' range
in the tropical Pacific Ocean.
        Giant clams are peculiar in that they rely for their nutrition on
symbiosis with a unicellular alga that transfers its products directly
to the clams. It is as if the giant clam were a plant itself,
extracting basic nutrients from the surrounding water. The clams can
be propagated quite readily, made to settle on rocks or slates, and
then set out their growing sites on the reef proper. They can be
farmed alone or together with other reef species (Table 6) at
intensities that will depend on how much management is applied to any
one reef site. The vast array of atolls in the Pacific Ocean and
elsewhere makes such reef polyculture another possible contributor to
the economic development of island nations. However, there are some
problems with the distribution of the products involved.
        Every year brings news of the closure of the lifecycle of yet
another marine food species; that is they are successfully bred, their
seed is reared, they are fed and brought to market. Compared to
species of domesticated mammals and birds which one can count on the
fingers of both hands, if not one, for the most prevalent ones, there
are many dozen species of fish and invertebrates which may one day
become domesticated. As control over the entire lifecycle of potential
cultivars is achieved one can apply techniques of modern genetics to
the upgrading of stock, such as selection for growth rate, fertility,
taste of the flesh, colour, and resistance to disease. Such controlled
breeding, even paired with some achievements of genetic engineering,

will transform what is still largely a craft into a true food industry
with a base in applied science.

TABLE 6. Commonly Marketable Marine Wild Crops that appear to be
Favourable on a Diversified Reef Farm.

Arthropods	Molluscs	Vertebrates
Mangrove crabs	Mussels - several species	Aquarium species
Penaeid shrimp	Clams	Dragons and lizard
	Oysters	fishes
Coalenterates	Pearl oysters	Groupers and other
Acroporoid corals	Tridacnids	fishes
Alcyonarian corals		Sea horses
Precious corals	**Seaweeds**	Sea snakes
Sea fans	Numerous colloid-	Turtles
	producing species	
Echinoderms		
Sea urchins		
Starfishes		
Trepang holothurians		

Note: Clearly, no one site would be suitable for all organisms in this
list, and many more could be added.

Source: Modified from M. Doty, "The Diversified Farming of Coral
Reefs", H.L. Lyons Arboretum Lecture, No. 11 (Honolulu, 1982).

     If one looks somewhat further into the future than the development
of biotechnology in aquaculture, which has already begun, one must
examine its possibilities in areas further offshore than where it is
now practiced. Increasing industry and tourism developments in many
coastal regions indeed suggest that one should at least examine the
potentials of disused oil platforms, for instance, or, further away in
time, that of open ocean OTEC installations. As to the former,
offshore oil platforms are rarely in calm seas and control of
surveillance of an intensity that would be necessary in aquaculture
would be most difficult, even if one were to grow molluscs that filter
the ocean water and need not be fed. Such practices could only pay in
fertile ocean surroundings where a gentle current would allow the
placing of banks of rafts for hanging cultures of shellfish or algae.
Such a combination of circumstances has not been demonstrated; one may
rightly doubt that it exists.
     If high seas, grazing OTEC plants for hydrogen or ammonia
manufacture were to have a place in the mix of energy supplies of a
more distant future, one may well speculate also about harnessing of
the large volume of their waste waters, more correctly the nutrients in
the deep water component of the discharge, to fertilize large areas of
the ocean. The discharge would be cooler than the tropical surface
waters in which the OTEC plants would be deployed and it would sink so
as to be removed from any control to influence the composition of its
food web. Even though some fertilizing effects might well occur, their

as to be removed from any control to influence the composition of its
food web. Even though some fertilizing effects might well occur, their
time course and location would be uncertain, as would be their
biological successions.

   Comparable obstacles would be met in the massive mariculture use
of waste water of shore based OTEC plants. The requisite of applying
control measures at many junctures of an aquaculture process speak for
the use of limited and measured amounts of OTEC waste water - if not
for a separate installation of deep water pipes, like at the Hawaii
Natural Energy Laboratory in Kona. If an OTEC plant were somewhat
offshore, as would be likely in some cases, because of bottom
configurations, large plastic retaining basins and associated
structures made of non-fouling material might well be deployed there,
as they might also be with future, open ocean OTECs. These would serve
the production of high value products with substantial technical and
material inputs.

   Not only for OTEC related mariculture but for all of it,
technical, energy, and material inputs determine what is cultured and
how much of it can be produced, the choice and levels of these inputs
are, in turn, influenced by market conditions, but in any case, the
bulk of mariculture will be found on or near the shore. Improvements
in materials technology and engineering that will advance aquaculture
will lie in containment and attachment devices with prominent attention
to antifouling properties, in new kinds of water conduits, and in the
control of water quality, especially with help of automatic
computerized recording devices. In the area of biological technology
one may look for biochemical measures to control stages in the life
cycles of the cultivars and for feed technology and sanitation. A
comparable assessment may be found in a listing of training subjects
for aquaculture in Southeast Asia, again compiled by SEAFDEC (Table 7).

TABLE 7. Ten training subjects required in the field of aquaculture,
listed by the priorities indicated by the Member Countries in Southeast
Asia.

Priority	Subject
1.	Fish Health Management
2.	Fish Nutrition
3.	Shrimp Hatchery and Nursery Management
4.	Marine Finfish Hatchery
5.	Culture of Natural Food Organisms
6.	Aquaculture Management
7.	Aquaculture Engineering
8.	Artemia Culture
9.	Sanitation and Culture of Tropical Bivalves
10.	Fish Fry Collection, Handling and Storage

what are complex biological and technical systems was shown to be the
crux of success in aquaculture so far; it is expected to continue to be
of paramount importance.

Availability of mariculture sites will be increasingly determined
by the resolution of competition with other coastal zone activities and
entitlement and other legal questions will gain in importance in this
competition. These aspects of aquaculture, as much as its technical
developments, will determine whether aquaculture can realize its very
great promise.

5.   ACKNOWLEDGEMENTS

Figures drawn by Laurel Lynn Indalecio.

6.   REFERENCES

Aquaculture - Moving from Craft to Industry (1988), Environment
    30(2).

Atlas of the Living Resources of the Sea (1981), Fisheries
    Department, Food and Agriculture Organization, Rome, Italy.

Cornell, I.I., ed., (1981) 'Advances in Fish Science and
    Technology', Fishing News Books Ltd., Farmham, Surrey, England.

FAO Yearbook (1988) 'Fishery Statistics, Catches and Landings',
    Vol. 62 (1986), Food and Agriculture Organization, Rome, Italy.

FAO Yearbook (1988) 'Fisheries, Statistics Commodities (1986)',
    Food and Agriculture Organization, Rome, Italy.

ICLARM Report (1987), International Center Living Aquatic
    Resources Management, Makati, Metro Manila, The Philippines.

Lloyd's Register of Shipping Statistical Tables (1988), London,
    England.

SEAFDEC Newsletter (1988), Vol. 1/2, Southeast Asian Fisheries
    Development Centre, Bangkok, Thailand.

FISHERIES ACOUSTICS:
ASSESSMENT OF FISH STOCKS AND OBSERVATION OF FISH BEHAVIOUR

R.B. MITSON
Ministry of Agriculture, Fisheries & Food
Directorate of Fisheries Research, Fisheries Laboratory
Lowestoft, Suffolk, NR33 OHT
U.K.

1.   INTRODUCTION

There is still an upward trend in the global yield from fisheries and
most of the 139 coastal states in the world need to know what fisheries
resources are available to them, although not all have declared
Exclusive Economic Zones (EEZ's). These 200-mile zones offer an
excellent chance to control exploitation and to manage natural
resources but there are difficulties in understanding the fluctuations
of fish populations and in forecasting their abundance.
      Fisheries were operated effectively all over the world for
centuries before the development of sonar, so this cannot be considered
as a basic need. The situation has, however, changed from one where
'all fishing is groping in the dark' to one where efficient detection
of many fish species has led to intense exploitation and depletion of
stocks. This in turn has placed great emphasis on the need to manage
stocks with care. Fortunately, technology, plus measurement techniques
and survey methods, have evolved for the purpose of estimating fish
abundance. The Food and Agriculture Organisation of the United Nations
(FAO) has played a major role in supporting such work in developing
countries. It has also been responsible for a number of relevant
publications of which a manual for aquatic biomass estimation
(Johannesson and Mitson, 1983) is the latest. A number of countries
participate in the Fisheries Acoustic Science and Technology (FAST)
working group which meets under the auspices of the International
Council for the Exploration of the Sea (ICES). The terms of reference
for the working group ensure that research effort is focussed on the
remaining problems.
      This paper outlines the technology and methods being applied to
the measurement of fish abundance and to the investigation of fish
behaviour.

D. A. Ardus and M. A. Champ (eds.), Ocean Resources, Vol. I, 251–258.

2.   ASSESSMENT OF FISH STOCKS

2.1. Accuracy and calibration of survey systems

There are two functions normally required of the fisheries echo-sounder/echo-integrator combination: (a) the ability to measure the target strength (TS) of individual fish; and (b) the measurement of volume reverberation from aggregations of fish. Knowledge of fish TS is essential for a quantitative measurement of biomass. It would be informative to examine the attainable level of accuracy and to compare this with the needs of fishery managers. Pope (1982) discussed this in relation to particular management goals. He showed that the overall accuracy of a survey to 95% confidence limits should be ± 3 dB (± 50%) for an exploratory survey, ± 2 dB (± 35%) for a time series of surveys calibrated against the virtual population analysis (VPA) method, and ± 1 dB (± 20%) when used as the only basis for setting the total allowable catch (TAC) from an exploited fishery.

The electronic equipment must be precisely calibrated to ± 0.5 dB and the terms of the relevant acoustic equation satisfied (i.e. the calibration factor, the mean TS for the fish species and, correction for energy losses due to the propagation path). The overall accuracy depends also on the sampling statistics of the survey track. In the echo-sounder the important parameters are the amplitude, duration and repetition rate of the transmitted pulse and the precision of the receiving time varied gain (TVG) amplifier which corrects for the range-related reduction in signal due to spreading and absorption losses. The signals must also be precisely demodulated and squared. The equivalent beam angle (Ψ) of the transducer is obtained from the actual beam pattern measurements in planes at $90°$ to the axis. This concept overcomes the problem of reduction in response away from the beam axis (within the equivalent beam response it is unity and outside it is zero).

The acoustic absorption coefficient (α) has been investigated in recent years (Francois and Garrison, 1982) but it is known only to ± 10^{-3} dB m^{-1} at 38 kHz, the most commonly used survey frequency, and ± 4 x 10^{-3} dB m^{-1} at 120 kHz. This factor sets a fundamental limit to the possible calibration accuracy; at 100 m range it cannot be better than ± 0.2 dB or ± 1 dB respectively.

Acoustic calibration must overcome the practical difficulties of far-field measurements at sea; the variability of hydrophones and, until recently, the inadequate standard targets. Compressional and shear wave speeds in copper spheres can be quantified to better than 0.1%, thus reducing errors in their TS to 0.02 dB. Calibration spheres of copper and tungsten carbide maintain TS to ± 0.1 dB from 0-30°C. The acoustic system can be calibrated overall at a selected range and signal level by locating a sphere, suspended on thin monofilament lines beneath the transducer. Once the sphere is located on the acoustic axis, its echo signal can be integrated over a period of time. Towed transducers have some advantages and are frequently employed for acoustic survey. Calibration is then more difficult to implement but a reciprocity method can be used with a similar order of precision to that of the standard target.

2.2. Fish Target Strength

Acoustic scattering from fish is considered to fall into the following regions:

(1) Rayleigh scattering - when the length of the fish is small compared to λ. Acoustic cross-section (σ) is proportional to the fish volume squared and to λ^{-4}.

(2) Resonance - which causes the swimbladder to vibrate at the frequency of the incident wave, thus absorbing and re-radiating more energy than predicted by the Rayleigh approximation.

(3) The transition region - where σ is proportional to the fish length squared and is largely independent of λ.

(4) Geometric scattering - when the fish length is much greater than λ. Note that the wavelengths most commonly used for fisheries surveys are 39 mm and 15 mm.

In situ methods of obtaining fish TS are preferred, but these pose problems of capture for species identification and measurement of the size distribution. The fish must be mono-specific, within a limited size distribution and be at a density low enough for the discrimination of single fish echoes. A case for further investigation of in situ TS of individual species is based on the use of all published in situ TS data on live fish to derive equations for use in 38 kHz echo-integrator surveys (Foote, 1987). By combining data for fish (such as cod, haddock and saithe) the result is TS = 20 log ℓ - 67.5 dB with a standard error (SE) of 2.3 dB probably due to differences in species and behaviour. For herring and sprat, etc. TS = 20 log ℓ - 71.9 dB, SE = 1 dB, where ℓ is the total fish length in centimetres.

A direct method of measuring fish TS in situ uses dual concentric circular beams (one wide and the other narrow). The main lobe of the narrow beam covers the same volume as the region of unity response in the centre of the wide beam. From the ratio of received echo strengths for the same fish in the calibrated system, the angle of the fish in the beam can be resolved and TS determined. Signals are processed via amplifiers with 40 log R + 2 α R characteristics for fish TS analysis and simultaneously in another channel with 20 log R + α R for echo-integration and hence total biomass estimation.

The split-beam principle is also used for in situ fish TS measurement. Here, the transducer is split into four quadrants for separate processing of the four sets of signals, combined in pairs. By measuring the phase differences between the port and starboard half-beams and the fore and aft axis the echo can then be compensated for the position of the fish in the beam and a TS directly obtained. During a survey, the criteria for single fish echo classification are critical; in practice finding part of a shoal where the fish density is low may be difficult but, when this occurs, it is possible to obtain TS distributions against numbers of fish. On the other hand, if the species is known, the TS distribution can be transformed to a size distribution.

2.3. Survey Methodology

The planning and execution of surveys must be carefully addressed if
the results are to be of value and allow comparison with other
countries results. One of the problems is detection and estimation
close to boundaries. Certain species spend a high proportion of their
time on the bottom then, in a short period, rise to a zone very close
to the surface. Both the near-bottom and near-surface zones complicate
the survey. The near-bottom zone, because of the need to bottom
reference the signals, plus the dead zone, which has a height of at
least half the transmitted pulse length (typically 0.75 m). For a fish
density of 50 m ^{3} the survey might miss 3.7 x 10^7 fish km ^{2}. To reduce
this problem, there is a need for wide-band transducers so that shorter
pulses may be used. Surveys in the near-surface zone need an upward
looking transducer but the unstable nature of the sea-surface and the
propagation losses due to high-density bubble populations cause
problems under some conditions.

2.4. Acoustic Noise

The most important single source of noise is the ship's propeller which
produces cavitation, causing high levels of noise at 38 kHz. It is
unlikely that many currently operated research vessels can survey at,
or close to, maximum speed because of this problem. A fisheries
research vessel propeller has to be a compromise in bollard pull, free-
running speed and low speed inception of cavitation. The effects of
noise on fish are hard to quantify. There is evidence of a high
sensitivity to low-frequency tones in the cod which has a threshold
close to ambient noise level in the range 30-470 Hz. Some ship
propulsion systems are using AC current generation which is converted
by silicon-controlled-rectifiers (SCR) to DC to drive the propulsion
motors. With 60 Hz generation and a 6-pulse rectifier, the ripple will
be at 360 Hz and its harmonics. This will be converted to vibration by
the propulsion motors, resulting in structure-borne noise radiated from
the hull as intense tones. Where there are strong peaks in the
frequency spectrum (e.g. between 200-700 Hz), tuna catches are
significantly less than those from similar vessels with a smooth
spectrum of underwater radiated noise. The offshore oil industry has
need to use high-powered sound sources in its exploration work and this
has prompted an investigation into the effects of airgun releases on
anchovy. First indications are that there are effects on some larval
stages and on adult fish.

2.5. Plankton Investigations

Predictions of the size and distribution of fish populations may be
hampered by the difficulty of making measurements on planktonic
organisms. To detect organisms between 50 μm and 10 mm in size,
fundamental research into their acoustic scattering properties was
necessary. An acoustic wavelength comparable to the size of the
organism is used to obtain a large echo. A multi-frequency acoustic
profiling system (MAPS) (Yuen, 1970) uses 21 discrete frequencies

logarithmically spaced between 100 kHz and 10 MHz. These are needed to
determine the numerical abundance, biovolume, or biomass, as a function
of organism size. Multiple plankton targets in the acoustically-
sampled volume at any given time mean that the method for extracting
size information differs from that used for fish sizing. The shape of
the scattering function and its dependence on both scatter size and the
acoustic frequency must be taken into account.

3.   OBSERVATIONS OF FISH BEHAVIOUR

3.1. Important Aspects of Behaviour

Effective systems and methods exist for the quantitative assessment of
many species of commercially-valuable fish. But knowledge of the
behaviour and distribution of these fish can be critical to success.
Much remains to be learned about the general pattern of temporal and
geographic distribution for many stocks. It is important to ascertain
if a particular stock can be identified, if it merges with others at
certain times, and to know the migration path in relation to the EEZ's
of countries expecting to manage and harvest that stock. The scale and
seasonality of population movements is fundamental to the effects on
fisheries of dumping wastes, marine pollution and the spread of
disease. Many stocks appear to be behaviourally, if not genetically,
distinct and make regular seasonal migrations along routes sometimes
determined by the patterns of tidal streams. Local movements often
seem to be diurnal and the behaviour of the fish can then restrict
their availability to fishing, or survey, because of concentration at
either extreme of the water column.

3.2. The Use of Acoustics to Study Behaviour

For many reasons, it is difficult to make optical/visual observations
of fish behaviour in the seas and oceans, but to obtain comparable
resolution to optics an acoustic system would need a focussing mirror,
or lens, with a linear dimension of 10,000 wavelengths. Considering
the energy losses underwater, a frequency of 1.5 MHz might be about the
maximum practicable, but even then the wavelength is 1,000 times
greater than that of light and the range would be only a few metres.
Most fisheries telemetry links have a restricted range (usually less
than 1 km). The limitation is due to restriction of both size and
weight of both the battery and transducer in order to minimise the tag
size. To achieve long range, high power and low frequency are needed,
but the size of a transducer is normally inversely proportional to the
frequency. Frequencies of between 70 and 300 kHz are most common, with
50 kHz being a typical lower limit.

3.3. High-information-rate Sonars

The closest approach to an optical system is probably high-speed,
within-pulse electronic sector-scanning sonar or, alternatively,
continuous transmission frequency modulated (CTFM) sonar. These

systems have the potential to detect and observe the movements of
individual fish, but scintillation of the echoes prevents individuals
from being identified with certainty. This makes it difficult to
ascribe any clear pattern of behaviour to an individual fish and
confirms the need for a means of positive identification.

3.4. Acoustic Tags and Telemetry Systems

Soon after the development of the transistor, the first fish tracking
by sonar of an acoustically-tagged salmon took place. Tagged skipjack
tuna (Katsuwonus pelamis) were tracked using a CTFM sonar which worked
at a frequency of 32-52 kHz in the search mode and 260-290 kHz in the
classify mode. The tag was a 50 kHz pinger placed in the stomach of a
tuna about 40 cm in length and its signals were received by the sonar
in the low-frequency listening mode (30-60 kHz). During the tracking,
it was possible to determine that the fish swam close to the surface at
night but that during the day it swam at various depths. Journeys of
25-106 km, made by the fish away from a particular bank at night with a
return each morning, implied that the skipjack tuna can navigate and
that it has a sense of time (Yuen, 1970).
    Electronic sector-scanning sonar has been used in a variety of
fisheries investigations. It works at 300 kHz and scans a 30° sector
at 10 kHz, transmitting 100 µs duration pulses at rates of 1, 2 or
4 s⁻¹. It is well suited to operate in conjunction with a fish-tag
transponder which comprises a 10 mm diameter cylinder, 45 mm long,
weighing 3.4 g in water. Attachment to the fish is by a thread tied to
a stainless-steel pin passed through the body of a flatfish or the
dorsal musculature of a roundfish. The sonar transducer can be rotated
to scan either in vertical (elevation) mode or horizontal (plan) mode,
so the transponder signal can be viewed in range and depth, or range
and bearing, relative to the ship's head.
    One of the first applications of sector-scanning was measurement
of the efficiency of a trawl used for the capture of flatfish. Clear
images of the component parts of the trawl, including the netting,
wires, otter boards, etc., made it an ideal tool for observing the
reaction of tagged fish to various parts of the trawl gear and,
ultimately, after a series of experiments, its efficiency in the
capture of plaice (Pleuronectes platessa L) was determined. It was
discovered that these supposedly bottom-living fish spend significant
periods of time in midwater and it became evident that they are using
what has become known as selective tidal stream transport. When the
tide flows in the direction in which they wish to travel, they leave
the bottom, find the zone of maximum water velocity, maintain their
heading and continue in this way until slack water, when they go to the
bottom to await the next favourable tide. This observation has shown
that the plaice in the Southern North Sea, when migrating, are only
available for capture approximately 50% of the time. It raises the
question of whether or not fish are carried along passively by the
tide, or if they swim to maintain their heading. To investigate this,
a compass tag has been developed. A pivoted ring magnet is used,
attached to a disc with a sector removed from its periphery. The
position of the sector is sensed by the one unbroken light beam of

eight. These are placed at 90° to the plane of the disc and
equidistantly spaced from one another. Interrogation from the sonar
results in a response of two pulses from the tag, the range from the
first pulse and the fish heading from the time period between the two
pulses. This design illustrates how a crude sensor can give useful
results.

   An adaptation of the same sonar and acoustic transponder permits
the real-time measurement of heart-rate from free-swimming fish by
using the heart beat of the fish as a timebase. As each beat occurs,
its electrical pulse is amplified and used as a trigger for the tag
transmitter. The resulting acoustic pulse travels to the ship whose
sonar is in the 'listening mode'. At the instant when the tag pulse is
received it causes the sonar transmitter to send a pulse which, on
arrival at the tag, interrogates it in the normal way. Thus, after
each heartbeat, the range and bearing of the tag are obtained, plus
other variables if required (e.g. the compass information). A recent
variation of this telemetry system employs a transponder tag
incorporating a sensor which measures tilt angle in the pitch plane of
the fish. This is used for the purpose of addressing the problem of
fish TS variation with tilt angle. An electrolytic-type sensor is
used, 11 mm long by 5 mm diameter, having a resolution of ±1°, over a
range of ±15°, and a measuring capability to ±50° (Storeton-West and
Mitson, 1988). Such telemetry systems are ideal for observing detailed
movements of positively identified free-swimming fish in relation to
fishing gear, bottom features, pollutants, tidal streams, etc., but
they have disadvantages. The most important is the relatively short
time of about 5 days continuous observation, which coincides with the
capacity of the tag battery and the typical endurance of scientific
staff. However, such observations are invaluable and are currently
unobtainable by other means. But the expense of the vessel, its staff
and the cost of maintaining the sonar imply that such a telemetry
system should be reserved for direct and detailed observations.

3.5. Future Developments

In areas where fish aggregations have a known, accessible distribution,
where they are also mono-specific and the fish TS is known, acoustic
surveys can provide quantitative results for management. However,
where there is significant mixing of species, the techniques currently
under development, which utilise wideband signals, may prove to be
valuable. For example, evidence is given of clear differences in the
frequency spectrum of such fish as herring, cod and saithe. There are
possibilities of using acoustic methods for in situ sizing of fish,
including resonance measurements and the Doppler structure. The
capability of acoustic systems to survey at greater depths can be
extended where necessary and better range resolution is also possible
to allow detection and estimation closer to boundaries, thus reducing
this source of potential error. Side-scan sonar can be used to detect
large fish over live reef areas, where conventional echo-sounding is
ineffective and nets cannot be used. The overall precision of fish
abundance surveys can be improved by obtaining a better knowledge of
variations in TS due to changes in fish attitude/orientation and

physiological factors. In the first instance, there is evidence to show that fish exhibit a wide range of reactions to noise stimuli which may lead to rapid movements and hence change of TS. Then there are the movements when feeding, some of which may lead to extreme body angles. Acoustic telemetry offers the best opportunity to make short-term, real-time, detailed studies of fish behaviour. But in order to answer many of the questions requiring longer series of observations, tags are needed with a capability to sense basic parameters such as temperature/depth/time relationships, to record these over several months and be capable of storing the information for several years. The challenge is in producing small sensors, adequate rates of sampling and long life in tags for the smaller, commercial species of fish. For these longer-term studies, we will need to leave acoustic telemetry and employ fish tags which have a large memory capacity from which data can be extracted after capture of the fish in the commercial fishery.

4.   REFERENCES

Foote, K.G. (1987) 'Fish target strengths for use in echo-integrator surveys`, J. Acoust. Soc. Am. 82 (3), pp.981-7.

Francois, R.E. and Garrison, G.R. (1982) 'Sound absorption based on ocean measurements`, Part II: Boric acid contribution and equation for total absorption, J. Acoust. Soc. Am. 72 (6), 1979-90.

Johannesson, K.A. and Mitson, R.B. (1983) 'Fisheries Acoustics`, A practical manual for aquatic biomass estimation, FAO Fish. Tech. Pap. (24), 249 pp.

Pope, J.G. (1982) 'User requirements for the precision and accuracy of of acoustic surveys`, International Council for the Exploration of the Sea, Copenhagen, Symp. Fish Acoustics, Bergen, No. 84, 9 pp. (unpublished).

Storeton-West, T.J. and Mitson, R.B. (1988) 'Fisheries telemetry technology`, pp.369-374 in Proc. Conf. Undersea Defence Technology, London, 26-28 October 1988, Microwave Exhibitions and Publishers Ltd., Tunbridge Wells, U.K.

Yuen, H.S.H. (1970) 'Behaviour of skipjack tuna, Katsowonus pelamis, as determined by tracking with ultrasonic devices`, J. Fish Res. Bd Can. 27, pp.2071-9.

PART VI

Space Utilization and Opportunities

THE OCEAN ENTERPRISE CONCEPT

DR. M.A. CHAMP[1]
DR. D.A. ROSS[2]
MR. C.E. McLAIN[3]
DR. J.E. DAILEY[4]

1. WHY OCEAN SPACE?

Today less than 1% of the annual resources consumed in the US comes
from the sea. Yet the Exclusive Economic Zone (EEZ) Proclamation gave
the US exclusive jurisdiction to the resources of the ocean out to 200
nautical miles. Making the US 1.3 parts land covered by seawater to
1.0 parts land (this is an addition of over 2 billion acres of new land
- doubling the size of the US). To date, the potentially great rewards
from the development of the resources of the ocean by the private
sector have been greatly inhibited by the scale of risks of such
candidate projects. The following areas show the most promise: marine
mining of coastal heavy minerals; ocean energy conversion; offshore
waste treatment plants (NIMBY); mariculture (fish and shellfish), and
platforms for air and space operations (floating ocean military bases).
   The construction of large (1-2 km^2) stable ocean platforms could
also provide this Nation with mobile overseas military bases to meet
the future need of decreased reliance on overseas military bases for
the USAF - Army/Marines - Navy, and for C^{3}I - Space Support - SDI. The
US has a worldwide military basing structure that will very likely
dwindle significantly in the next ten years. Air bases in Panama,
Spain, and the Philippines are becoming extremely expensive and less
useful and less available . The Soviets have approached the problem
differently by employing mostly movable or removable assets (floating
piers, tenders and repair ships, floating drydocks). Ocean bases can
provide key aspects to hemispheric defense systems including border and
internal defense. Large stable ocean bases can also serve as Centers
to suppress sabotage, terrorism, and narcotics trafficking, arms
shipment; major weather stations for enhanced weather prediction,
global climate studies, air traffic routing centers (considerable fuel
savings); alternative energy generating plants (OTEC); and serve as
platforms to provide indirect US military assistance to third world
countries (such as training, intelligence, communications,
transportation, construction, medical supplies, physicians or disaster
relief, logistics, etc.).

[1] National Science Foundation, 1800 G Street N.W., Washington DC 20550
[2] Woods Hole Oceanographic Institution
[3] Consulting Services
[4] Brown & Root Inc.

D. A. Ardus and M. A. Champ (eds.), Ocean Resources, Vol. I, 261–274.
© *1990 Kluwer Academic Publishers. Printed in the Netherlands.*

The Ocean Enterprise Concept has been proposed as an exciting and
challenging mechanism for launching a new era of awareness, practical
development, and utilization of ocean resources beginning in the early
1990's. It is only through a cooperative "pulling together" of
government, academia, and industry, that significant new areas of
operational economic interest can be developed or current ones strongly
bolstered, in the oceans sector. The original "Stratton Report" (Our
Nation and the Sea, 1969) recognized the great basic potential of the
oceans and provided a broad discussion of the many appropriate areas
for scientific and economic development. As assessment at this time
(some 20 years later) suggests some interesting observations:

-    Great strides in the scientific understanding of the oceans have
     been made in some areas (the recent work on rift zone geology,
     thermal vents, and their implications for ocean chemistry and
     biology, for example.

-    No new major economic area has been developed in the ocean sector.
     The principal economic payoff areas remain those of: shipping
     (merchant marine), fisheries, and offshore oil and gas. Heavy R&D
     investment has been made in such areas as mineral deposits
     (manganese nodules) and OTEC, but no practical business of net
     economic value has developed.

-    A strong well recognized constituency has not yet developed for
     the oceans, although a lively basis for such a constituency
     appears to exist.

Many actions have been initiated and ideas and technologies
developed which, if supported under a strong long-term commitment by
the government, academia, and industrial sectors could provide a basis
for very significant scientific and economic expression of our use of
the oceans.

2.   THE PRIVATE SECTOR ENVIRONMENT

Perhaps the major "disappointment" of the past 20 years has been the
failure of any major new ocean economic area to develop. Broad
technological and economic constraints have been suggested as the
primary factors in preventing many of the Stratton Report goals from
being achieved. The development of these ocean resources (from the use
of ocean space to the development of individual resources) has been
constrained by the lack of: public/private venture infrastructure;
legal/regulatory implementation strategies; environmental, economic,
social, and political; and technical and engineering problems that
arise from the "marinization" of land based engineering concepts,
technologies, structures and facilities for use in ocean enterprises.
The limiting factors are really leadership, infrastructure and venture
capital (because the scales of risk are perceived to be large). The
infrastructure needs can be developed and supported by a Federal-in-

house incubator, an ocean going Fannie Mae, and a quasi government non-profit corporation (chartered through federal enabling legislation). This "quasi" government corporation is needed to: provide the limitation of liability to that normally accepted by the Federal government, and minimize the risk of intervenor legal action (similar to the Trans-Alaskan Pipeline, or COMSAT Corp.). Several national and international workshops have stated that they key technologies exist, but have not been utilized in such a manner on a commercial scale.

It is in the interest of this Nation to create organizational infrastructures which bring together the resources of government, industry, and the academic sectors to undertake large scale resource and technology development projects, where scales of time, risk, and/or magnitudes are too great for one sector to go along to bridge the NO MANS LAND gap between research and development as illustrated in figure 1. The bridging mechanism requires a larger more integrated effort with private/public sharing of funding to support special development activities.

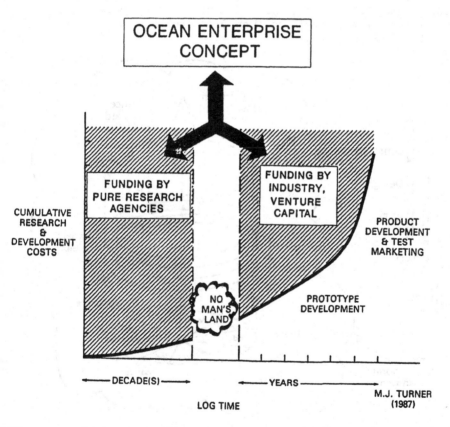

Figure 1. The ocean enterprise concept to bridge the NO MANS LAND gap.

3.   WHAT IS AN OCEAN ENTERPRISE?

Ideally an ocean enterprise should: generate economic revenues, not
only by cost savings but also generate new revenues; create jobs and
economic benefits from the development of ocean resources, and
technologies; protect and conserve the developed resources and ocean
environments; provide the benefits of a public service, and reduce
public risk; and support the Nation's interests overseas.
      It is also timely and desirable to foster new civilian and
military partnerships to enhance this nations competitiveness and
economic growth. With current and future budget limitations (the trade
deficit and the increasing national debt). It is also desirable to
stimulate military/civilian synergy, because projects of this scale
(such as large ocean platforms) have costs that require multiple use
benefits to society. Also these large projects must maximize the
commercial spin-offs (e.g. as NASA with the space program) to increase
the benefits to society and distribute the construction and operation
costs across a wider array of users. Different kinds of ocean
platforms (from ocean airports, mining facilities, to recreational
facilities - hotels, resorts, etc.) will spin-off entirely new
commercial industries, providing significant public and private
economic benefits (Fig. 2).

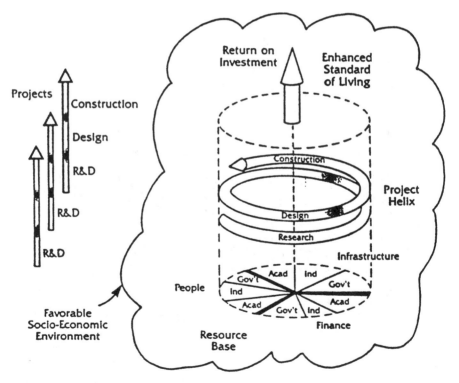

Figure 2. Partnership of industry, academia, and government to conduct
research, design, and construct large scale ocean projects.

Ocean enterprises have to develop from a succession of smaller scale projects that perhaps develop for application in the shallower, near coastal waters, providing local public service benefits. These projects should have a dual use being initially developed with public-private sector funds for civilian use, however, engineered and evaluated with a military perspective and application in mind. Examples of these could be a moored floating ocean platform designed and engineered for a NIMBY public service project such as a coastal airport or waste treatment facility (e.g. high temperature garbage processing and treatment plant). In the US not a single major airport has been built since the early 60s. By the year 2000, 80% of the US population will live within 60 miles of the coast, and public air transportation which has air space limitations today, will not be able to meet the demands for services in large US coastal cities (New York, Los Angeles, San Francisco, etc.). The construction of stable ocean platforms for airports could be initially supported by public and private funding with repayment from user fees and capitalization of infrastructure.

Vigorous efforts have been supported by the US government, to encourage private capital involvement through a "cooperative partnership" approach to the development of new technologies and to the transfer of technology from the R&D environment to practical economic application in the marketplace. These have, in some cases, shown a modest degree of success. Yet, significant investment from the private sector for potential new economic areas is unusually lacking. "Why doesn't industry get more excited about the oceans?", Federal agencies and academia have asked. The direct answer is that there are no perceived returns on investment justifying the perceived risk.

4.   FUNDING LIMITATIONS

Due to the increasing National debt and the trade deficit, it will be the policy of the New Administration to reduce Federal support for many programs for which such support would seem to be reasonable and appropriate from the state and local government and/or private sectors. Thus, under the new Administration, one cannot really expect major direct Federal financial support for the increased efforts needed to truly develop and exploit new ocean resources. It can also be argued that, even under a variety of administrations with differing philosophies, there has really never been a basis for a major increase in Federal funding for major new ocean projects. The constant dollar support for underlying programs in basic and applied ocean sciences, charting and mapping, and related activity has been maintained at a fairly steady rate (though slowly decreasing through inflation). This situation is not likely to change in the future, regardless of administration. The real question is: with the understanding that major increases in developmental funding from the Federal Government or from foundations are not to be expected, is there any basis for a major increase from the private sector?

5.   PRIVATE SECTOR INCENTIVES FOR OCEANS DEVELOPMENT

The prime requisite for private capital financing of development is
that an investment yield a profitable return. Moreover, most investors
or companies looking at rewards for their own R&D efforts, tend to
evaluate the worth of investment opportunities in terms of near-term
returns. The combination of potential immediate tax benefits and
profitable return on investment realized in the 2-5 year time frame
seems to be characteristic of the most attractive private sector
investment opportunities. Investments in 5-20 year time payoff areas
are only of interest in certain highly specialized industrial areas:
oil and minerals exploration, timber production, and development of
public and private power resources, to name a few. These long term
areas are characterized by the long lifetime of assets and the highly
predictable long-term general requirements for their products (although
short term market conditions may fluctuate wildly). The basic reason
that short-term payoff is so important is that futures can be
"reasonably" predicted over a 2-3 year period at most, in the
perceptions of a majority of investors.
     What constitutes an attractive R&D investment opportunity in the
private sector? A few critical elements which must be present to
validate a "first class" investment risk are suggested below:

-     A well defined end product.

-     A well prepared development strategy and business plan.

-     Definable markets and specific paths to those markets.

-     A basis for prediction of "comfortable" profit margins.

-     Technological uncertainties well defined and directly addressed by
      a planned R&D program.

-     The business plan adequately structured within the international
      and national socio-political and economic environment.

-     A direct identified path for Return on Investment (ROI).

This implies a clear economic "model" for the investment program
showing how ROI will be generated.
     There are doubtless additional important points. The bottom line
argument is that the level of private investment will be strongly
coupled to the degree to which the investment environment and market
place is understood, and the potential for a reasonable ROI.
     It seems self evident that the reason for capital investment in
the oil/gas, merchant marine, and fisheries areas, basically lies in
the fact that these are perceived as well understood economic areas by
the operators and investors. In the oil/gas area, certain
developmental investments are regarded as essential, based on past
experience (the need to explore, develop more efficient techniques,
etc.). In shipping and in fisheries, requirements for investment in

capital equipment are well understood, but the potential of R&D to
improve profits through increased efficiency, understanding, etc. is
less well recognized or accepted. New techniques in these traditional
areas often must be spurred by Federal government R&D or by regulation.
    In most other "new technology" areas of ocean exploration,
development has been all but non-existent outside of those projects
based on Federal government funding. A notable exception has been the
once vigorous but now moribund investment in the manganese nodule
mining potential. Here, it looked as though all elements were in place
to make industrial investment attractive, and huge amounts of
investment were actually undertaken by several large consortia of
companies. As it has turned out, a declining metals market, the
general long-term malaise of the world economy, and the recently
concluded Law of the Sea Treaty, have all acted together to make the
planned manganese nodule industrial development economically
unfeasible. This has stung the investors severely, and has contributed
to a doubly cautious approach on the part of private investors with
regard to future opportunities requiring major investment levels.

6.   AN APPROACH FOR THE OCEAN ENTERPRISE CONCEPT

An "Improved Approach" would seem to be a necessary part of any program
designed to make the Ocean Space Initiative a success in terms of the
infrastructure that will lead to economic benefits. In addition to
activities well within the operational potential of the current oceanic
community, a new set of techniques and organizational methods must be
developed. These methods must be expected to enhance the probabilities
that major projects and new economic potentials will in fact be
realized, as opposed to just being studied, evaluated, and then left to
await some future development. This "Improved Approach" must in fact
concentrate on bringing new elements of the private sector strongly
into the ocean development arena. Government and academic resources
are already there and do not have major new sources of support to draw
from. Under this "Improved Approach" some things need to be recognized
as rather fundamental:

-    The basic incentive for private sector interest is that of
     perceived future return on investment.

-    The basic reality for the legislative branch of the government is
     the perceived connection between congressional action and the
     reaction of individual congressional constituencies (i.e. programs
     must have real social/political/economic impact on real
     constituencies of Senators and Representatives). Strong private
     sector involvement enhances legislative interest.

-    The primary resource for basic ocean research lie within academia
     (including the various oceanographic institutes) and the Federal
     Government.

-   The primary resources for development lie within the private
    sector.

-   The private sector is becoming increasingly aware of the perils of
    investment posed by uncertainties in policy and the socio-
    political environment. A carefully planned and prepared basis of
    support in these areas, as well as a generally favourable economic
    projection, is becoming a requisite to investment.

Considering all of the above, a series of new mechanisms (as a means of
attracting new private sector participation) is suggested to introduce
new programs and initiatives.

7.   "TRIPLE ALLIANCE" R&D PARTNERSHIPS

This mechanism is used to develop a model for R&D Limited Partnership
investments. It proposes a partnership of government, academia, and
private capital sectors to establish long term and vigorous support for
applied R&D in the oceans area. The proposed structure is diagrammed
in figure 3(a). An operating R&D center established under the auspices
of one or more academic institutions. Support from the private sector
is provided through an R&D Limited Partnership (RDLP), thereby
providing a direct path for technology transfer and market application
directed by the General Partners through agreements with "user"
industries, who pay royalties to the RDLP in return for manufacturing
and marketing rights received. The royalties are used to provide
return on investment to the RDLP investors, and to self-endow the
center after the RDLP limited partners are paid out. Ideally, the
academic participants will be one of the General Partner. Figure 3(b)
is a more complex example of a 2 phase model.
    Government and non-profit foundation support is separately
solicited by each project, but has the added attractiveness for
projects and grants by providing "leverage" based on the concurrent
programs supported by the RDLP contracts with the Ocean Space
Initiative. In this way, a formal tie is established between the
successful market application of project developments and the future
financial support of the initiative itself. Industrial (private
sector) participants are protected by limited investment and benefit
both through tax credits and by RDLP distributions or individual "user"
contracts. The General Partners, through the RDLP, provide sufficient
isolation that antitrust requirements are met. Finally, joint academic
and government support of the center helps validate the products with
are applied and sold through the private sector channels. This
validation can be critical in reducing perceived risk and encouraging
private sector investment. This mechanism can provide a helpful
"umbrella" for encouraging a significant increase in private sector
investment.

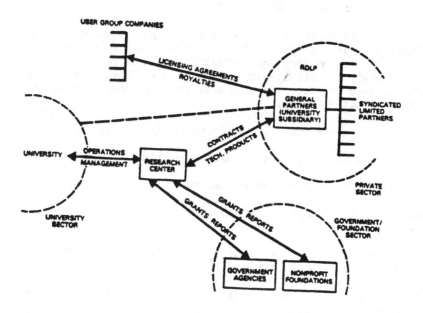

SUPPORT OF A NEW ECONOMIC SECTOR DEVELOPMENT

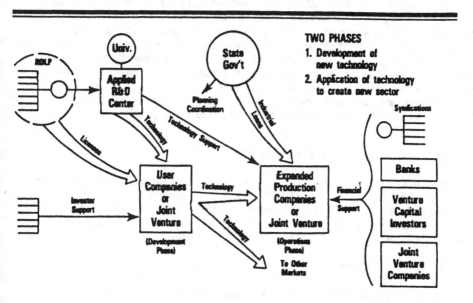

Figure 3.   Simple (a) and Complex (b) examples of organizational infrastructure models for ocean enterprise projects.

8.   MAJOR ECONOMIC AREA JOINT VENTURE

This mechanism also emphasizes the role of the private sector.   Rather
than relying on a single major industrial developer, a team of
industrial and investor partners, perhaps through a joint venture
corporation, would be established to develop a particular area of great
potential economic benefit.   Such an area would be characterized by
some of the following traits:

-    Development path featuring a graded investment startup.

-    Backup by favorable policy and socio-political environment.

-    Meets environmental protection issues.

-    Takes short and long range economic and market conditions into
     account.

-    Does not require development of basic scientific understanding,
     i.e. is a technology development.

-    Scaling models exist for transition from laboratory to industrial
     practice.

-    "Spin-off" developments are inherent in the approach.

These fundamental traits allow the development of a business plan which
permits the highest technological risk problems to be solved with a
modest initial investment.   Operations and market testing would be
conducted under a prototype operation which again does not require a
full scale manufacturing investment.   It is also important to identify
multiple potential paths for market development so that more than one
option for investment payback exists.   A goal for the Ocean Space
Initiative might be to launch in early 1990 at least one major effort,
with an initial 10 year development and business plan.

9.   NATIONAL OCEAN POLICY STATEMENT

Simply stated, this would seek to provide an Ocean Economic Development
Policy statement by the President which follows up on the EEZ
declaration of March 1983 and would include specific backup actions
which strongly support ocean resource development by the private
sector.   Features of the policy should include:

-    Operation within the EEZ.

-    Cooperative guidelines for working with LOS signatories.

-    Policies providing incentives for ocean development investment
     (tax credits, small business loans, SEC and Justice opinions on
     consortia, etc.).

- Emphasis on ties to economically depressed areas.

- Assurances on commitments to protect ocean environment while undertaking development, i.e. an intelligent balance between ecological concerns.

10. COALITION OF OCEAN RESOURCE STATES

In conjunction with those new incentives and special projects which could be directly encouraged by Federal Government actions, a movement to organize those states having direct or strong indirect interests in ocean development as well as currently established ocean business would be helpful in establishing a strong constituency for the oceans. Similarly, a parallel association of city governments might be developed for ports and other cities whose economic basis may depend or could depend heavily on ocean development and economies. Spearheading this effort should be the ocean related industries and professional societies, backed up by the general interest of the Department of Commerce, National Science Foundation, Transportation, and Interior. Such an organization would, through the associated Congressional delegation of the member states, have a strong effect on the development of a broader Congressional constituency as well. The coalition would, among its goals, act to:

- Promote understanding of the interdependence of state and local economy and environment on ocean related development and industry, both present and future.

- Identify and support the development and adoption of appropriate policies, legislation, and planning at local, state, and national levels which will best serve state and local requirements and interests, as well as meeting national concerns.

- Develop appropriate state-to-state operating relationships and agreements which will aid in the beneficial development of mutually shared ocean related opportunities.

    One very appropriate path to the establishment of such a coalition may well be to obtain the support of the following organizations: Coastal States Organization (CSO) [the sponsor of the Ocean Enterprise Workshop with the Marine Technology Society at the Ocean '88 Meeting in Baltimore, Maryland, November 1, 1988]; the National Association of Counties (NACO); the US Conference of Mayors (USCM); and the League of Cities (LOC). These groups already have well knit operating committees and organizational objectives which broadly parallel the actions suggested above. For example, among the NGA standing committees, one or more of the committees on: National Resource and Environmental Management; Transportation, Commerce and Technology; and Community and Economic Growth, might be very receptive to developing a working group of interested states in the ocean area. Further, the NGA already sponsors various coalition (cf: Coalition of Northeastern Governors) and Regional Commissions (cf: Four Corners Regional Commission, Old West Regional Commission, etc.). Such organizational sub-elements are a natural part of the NGA operating structure. The USCM, LOC, and NACO have similar objectives, structure, and operating methods. USCM maintains continuing efforts in areas of Energy and Environment, Urban Economic Policy, and Transportation, for example. The development of such a coalition in formal recognition of the increasing importance of the oceans as a major factor in state and local economic structure, would provide a strong and effective boost to the Ocean Enterprise Concept.

11. IMPLEMENTATION STRATEGIES FOR THE OCEAN ENTERPRISE CONCEPT

The overall program for Ocean Enterprise Concept should, of course, embody much more than such major new thrusts as are discussed in the proceeding sections. Implementation of the Ocean Enterprise Concept ought to establish a total environment for the enhancement of ocean related activities and interests of all types. The principal measure of the long range effectiveness of the program will be the initiation of major new development areas which can sustain growth. Without the total environment created by the program such "new approach" initiatives as are discussed in the previous section would have little chance of successfully developing. Without the resulting realization of such new initiatives, the program would be judged, over the long term, as a failure. The program must then have two objectives which are interrelated:

-    The creation of a heightened environment of ocean related awareness and actions.

-    The initiation of some significant new development with both technological and economic impact which will last (i.e. become an integral part of the national dynamic economic structure).

If these two objectives are met, then the Ocean Enterprise Concept may well be judged to have ushered in a new area of ocean utilization.

The foregoing arguments suggest that the basic operating approach for the program must be that of a team effort, with coordination by key Federal Agencies backed by specific White House approval, supporting major activities in five areas:

-    Policy development.

-    Constituency establishment.

-    Awareness enhancement.

-    New enterprise initiation.

-    Research and development direction and augmentation.

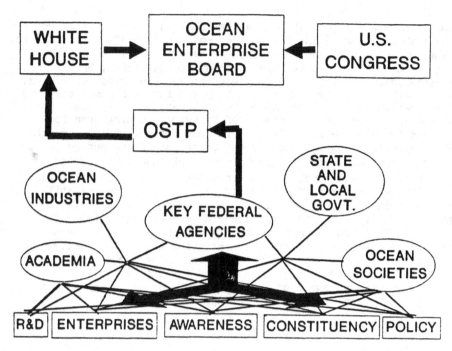

Figure 4.   Proposed national team effort for the Ocean Enterprise
Concept.

Figure 4 suggests a general way in which the team effort might
take place under Federal coordination. Each of the participating
sectors would contribute to the appropriate activities through a
program master plan. Participation by the White House (particularly
the Office of Science & Technology Policy) and other departments and
agencies would be most important. Especially beneficial would be a
Presidential memorandum or statement ushering in the program, and
designating Federal agency responsibilities. Ideal timing for such a
statement would be February 1989.

A large number of special national, regional, and local activities
primarily stressing ocean awareness and education should be encouraged
throughout the US. The professional and special interest societies
should be encouraged to sponsor special appropriate events as the
keystone to a broad basis of public participation. Special materials
could be provided for the use of school systems and by national
resource management agencies of state and local governments. Through
the USCM, LOC, and NACO special Ocean Enterprise Workshops for local
governments could be sponsored.

It is certainly apparent that there is not much lead time for
planning. For many of the activities there is only time for a short
preliminary "warm-up" period prior to the actual activities of the
Ocean Space Initiative. Thus, if any success is to be based on
adequate preparation, that preparation should start quickly. To
achieve full operating capacity in Spring of 1989, many of the

projects, committees, educational activities, and the like will require
start-up efforts during the first three months or so of 1989. Finally,
there are many varied functions subsumed but not listed under the
"Formal Program" line. Literally hundreds of individual, but
interconnected, activities and projects can and should be undertaken.
These are all "to be developed", first by the informal committee, and
then by a whole fabric of groups working at national, regional, and
local areas. The early development of a larger planning committee
would seem to be essential. This "full" committee would prepare the
detailed operating plans for the Ocean Space Initiative and review its
operation. Many more special events may be planned and expected.
After all is said and done, however, the Ocean Space Initiative will
receive its best marks in future historical reviews from its ability to
provide a springboard for even one or two major new ocean related
programs or activities with significant and long lasting impact. The
timing is well suited to capture the highest probability of such
success, with credit to the new Administration but with potential long-
term rewards transcending any administration. The technology, ideas,
and interest are present. The Ocean Space Initiative can, if
vigorously encouraged and coordinated, yield a strong turning point for
the history of ocean research and enterprise.

FUTURE OCEAN SPACE UTILIZATION IN JAPAN

KENJI OKAMURA
Special Assistant to the Minister of State
for Science and Technology,
JAPAN.

1. NATURAL FEATURES OF THE ISLAND

Japanese islands consist of 6,852 islands with a total area of
377,303 km² and with a total length of coastline of 33,037 km. Nearly
46% of this is designated as the natural park coastline. As for the
major four big islands which have the main industrial and economic
activities, nearly 50% of the total coastline (18,700 km) is artificial
coastline, where reclamations, bank protections or other artificial
treatments have already been applied. Another 25% is natural coastline
with bluff and the remaining 25% is natural coastline with shoaling
beach with grits or mud. Two thirds of the total area is mountainous
and is not convenient to live or work in, and one third is the
habitable area including working and farming areas. Reviewing old
history in this country, fisheries have been quite popular from the
early days and fishery products have become deeply engrained in the
diet of Japanese people, because of such favourable oceanic conditions.
On the other hand, marine transportation had been the most important
means for communication with other countries and also for domestic use.
A construction work of man-made island covering 30 hectares of the area
with 1.4 million m³ of materials was performed during the years from
1180 to 1196 at the harbour of Ohwada (old Kobe) for the purpose of
getting a calm area protected from the strong wind from the west.

2. DISTRIBUTION OF THE POPULATION

Until the middle of the 19th century, the increase of the population of
Japan had been limited by the national self-supporting and self-
sufficient policy during the period of seclusion. The population in
the year of 1858, when the national isolation was changed, was nearly
30 million. After opening the country to the world, the population has
increased to 121.049 million and it is anticipated to be 131.192
million in the year 2000 and possibly it might be saturated at around
140 million in the early half of the 21st century.

Among the major industrial nations, Japan is uniquely lacking in
natural resources, except some marine living resources. Many raw

D. A. Ardus and M. A. Champ (eds.), Ocean Resources, Vol. I, 275–282.

materials and energy resources have to be imported and the products
exported worldwide to support our lives. Therefore the marine
transportation is significant for such import and export. Many
industries requiring those raw materials had to be located in the
coastal zones, where the most efficient operation was expected without
any additional domestic transportation cost. Nowadays, almost all
countries in the world can not maintain their modern standards of
living without obtaining the resources and energy through international
trade.

Out of the largest ten cities (over one million in population) in
Japan, eight cities are located facing the coastline, the other two are
in inland areas. According to the result of a survey on the
populations of 100 large cities, it is reported that nearly 72% of the
population is residing in the cities located facing the coastlines.
Population and industries are always seeking closer location to the
coast to assure their flourishing circumstances. Similar tendency will
be found in other ocean-bounded developed nations. These situations
were not caused by personal preference of each individual, but by
economic and social structure. In the recent world, an information-
intensive society, the instant transmission of information has become
possible by means of electronic media, including telephone, facsimile
and information network system. However, this instant transmission of
information will not always solve the problem of concentration of
population to the big cities, but sometimes causes needs for a faster
transportation of persons, or need to live within a shorter distance.

According to a recent study on the expected variations of the
numbers of persons engaged in various industrial fields in the year
2000, we foresee a great change in the structure of society as shown in
Table 1.

TABLE 1. Numbers of persons engaged in various industries

Unit: million.

Year	1985	2000
(1) Primary industries; Agriculture, Forestry, Fishing & mining	5.18	3.50
(2) Manufacturing of Materials: Basic metal, Minerals, Pulp, Paper, Chemicals, Textiles, Pet. & Coal products	3.20	2.28
(3) Manufacturing of Machineries: Fabricated metal products, Machineries, Electrical machineries & equipment, Transportation equipment, Precision instrument	16.19	16.47
(4) Manufacturing of Food and Beverages & others	5.14	4.60
(5) Construction	5.30	6.14
(6) Electricity, Gas & Water supply	0.33	0.34
(7) Transportation & Communication	3.43	3.29
(8) Wholesales & Retail Trade	10.78	10.79
(9) Finance, Insurance & Real Estate	2.17	2.60
(10) Management and Information service	3.66	6.63
(11) Medical service	2.11	3.19
(12) Education service	1.98	2.32
(13) Recreation and Culture service	3.81	5.41
(14) Households and Personal service	1.80	2.27
(15) Public Administration & others	2.99	3.28
Total	58.07	63.11

Besides the above problem, this country has to depend heavily on the import of food, raw materials, energy and other products from many foreign countries. Marine transportation is still one of the most important points. In addition to the foreign trades (778 million tons annually), the figure of domestic marine transportation (210 billions ton/km in weight multiplied by distance) corresponds to nearly 50% of the total domestic transportation including trucks and railroads. There are nearly 1100 harbours with cargo handling facilities and also more than 2900 harbours for fisheries. These harbours occupy the coastal sea areas of 10,670 km². Taking account of fishing activities in the coastal waters, it is clear that we should consider utilizing the ocean space deeper than 20 m in the near future.

During the past 40 years, the world fishery harvest has grown at a good pace and the present amount is approximately 75 million tons per year. Already some 15% of the world food supply of animal proteins come from fisheries and the figure in Japan is as high as 45%. Japan is faced with a need to meet rising and diversifying demand for marine products and to increase fishery resources in neighbouring waters in order to maintain a high degree of self-sufficiency. Therefore we are trying to increase the catches from our coastal and offshore areas applying more advanced technologies in aquaculture and fish farming. The fish farming is a popular subject and the culture and rearing of marine organisms in certain localities will be promising.

As an example of near-shore space utilization in olden times, reclaimed lands by drainage for farming had been completed in the 7th century and some large scale reclamation works were taking place in old Osaka and old Tokyo, areas of 120 and 180 km² respectively, at the beginning of the 17th century for the purpose of urban development. The land area reclaimed along coastlines is nearly 1556 km² in total so far, including 21 man-made islands built during the last thirty years. The area of these islands is 72.8 km² in total, built at locations of shallow water depth less than 20 m in bays or in inland seas. The islands have their own main purposes such as ten harbour facilities, two airports, two industrial lands, two oil storage facilities, two waste disposals, one power plant, one coastal park and one coal mine. Some islands are multi-purpose. For example, seven of the above had aimed at urban development combined with harbour facilities. Sometimes these islands were planned to secure their sites for the facilities in order to reinforce the social activities on land. These transfers of the facilities into sea areas may cause change of environmental condition, bringing about serious conflicts with fisheries cooperatives and regional residents. We should be careful of the environmental effects of any artificial construction in coastal waters, giving deliberate scientific investigation on the environment without any underestimation or unreliable exaggeration.

3.   RECENT CONSTRUCTION OF MAN-MADE ISLANDS

During several years in the '80s, a number of large scale man-made islands in water more than 10 m deep were completed such as Kobe port Island (436 ha), Osaka South Port (937 ha), Yokohama Daikoku (321 ha),

Rokko Island (583 ha) and Yokkaichi (387 ha). These have brought
great benefit and have given a stimulus to activate each surrounding
zone. However all the cases were in near-shore, not in offshore areas.
It is considered that offshore man-made islands will give us another
different opportunity to utilize wider sea areas than near-shore, for
example the sea area between the island and the coast will be a calmed
area very desirable for recreation or aquaculture or other purposes.
     The construction of the largest man-made island offshore, the
Kansai International Airport, started in 1986 and is now underway. As
the first phase plan, the airport will be opened in 1993. The policy
of this airport is considered as follows:

-    both international and domestic air transport,
-    24 hour operations,
-    environmental protection, in water and in air (noise)
-    possible further expansion.

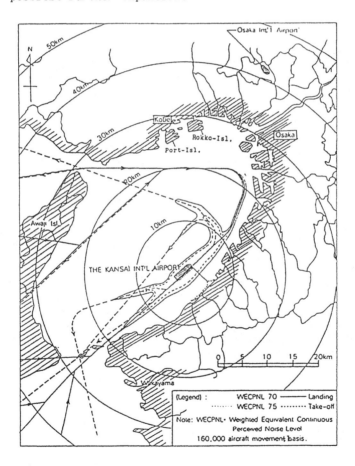

Figure 1. Noise exposure forecast contour map.

Figure 1 shows the map of the surrounding area and the noise
exposure forecast contour. The contours of Weightened Equivalent
Continuous Perceived Noise Level (WECPNL) 75 do not overlap the land
area, assuring full protection of the land environment from noise
pollution. The 24 hour operations in this airport depend heavily on
this assurance. The total area of the land will be 511 ha, surrounded
with a seawall 11 km in length. Airport access will be a road and rail
access bridge, approximately 4 km long to connect the airport island to
the mainland. In addition, the use of high speed boats will be
available, providing a calmed sea access area protected by a
breakwater.

In Phase I, one runway of 3500 m length will be completed, with a
capacity of nearly 160,000 take-offs and landings per year. The
estimated number of passengers will be about 25 million per year. The
volume of cargo will be more than one million tons, which is comparable
to Narita airport, the largest cargo handling airport in the world.

With 24 hour operational capacity, this airport will be able to
provide scheduling flexibility to airlines thus allowing for more
character and extra flights, both passenger and cargo, which are
presently under tight control. This airport will be particularly
popular with the import and export business, because they favour
freight movement and customs clearance at night to avoid overnight
storage and loss of time.

Figure 2 shows the future expansion plan to meet the increase in
demand, providing three runways.

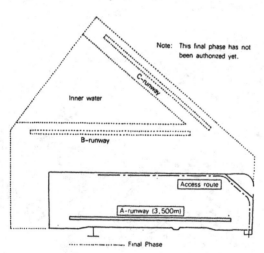

Figure 2. Future expansion plan.

The average water depth at this site is about 18.5 m and bottom
soil includes an alluvial clay layer deposited during the Holocene.
This clay layer, after being improved, will be compressed by the weight
of an approximately 30 m thick layer of reclaimed soil so as to set the
clay before the opening of the airport. New excavations are necessary
to supply the 150 million m³ of reclaimed soil. Furthermore a

continuous supply of a large quantity of reclaimed sand and stone is
also required. An elaborate computerized system will be used for
precise process control of this reclamation work, which is of
unprecedented dimensions. Water quality and other conditions will be
monitored during the construction to keep adverse environmental effects
to a minimum.

The reclamation works are being conducted at many corners of the
coastline. Some other large scale typical projects are the man-made
islands of Tokyo City in Tokyo Bay and the transverse tunnel and bridge
crossing Tokyo Bay including the man-made islands between Kawasaki City
and Kisarazu City.

4.   PLANS FOR OCEAN SPACE UTILIZATION IN THE FUTURE

As for the utilization of ocean space, there will be subjects in two
different categories.

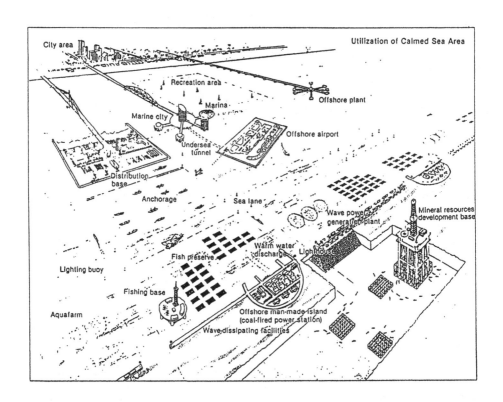

Figure 3.   An example of the utilization of a calmed sea area.

One category is to secure the space needed for construction of better urban structure with more convenient communication systems to support the future prosperity of the zone. This is needed because the existing high density utilization of the limited area on land has brought about many social as well as economic problems incidental to the overcrowded urbanization.

The other category is the utilization of ocean space with a specific object of each local interest concerned such as airport, port and harbour, fisheries and the products, recreation, marine research facility etc. Many advanced technologies will be applied for extraction of wave power, utilization of deep water to boost the basic biological production, current control, artificial upwelling, various marine structures and facilities, monitoring of sea farm and environment, etc. Figure 3 is a conceptual example of the utilization of a calmed sea area.

The governmental Agencies are taking initiatives to cooperate with many societies and sectors to study the utilization of ocean space as follows:

Agency	Project	
S & T Agency	Aquamarine	Selection and implementation of development for technologies
National Land Agency	Marinopolis	Regional development through use of resources by marine industries
MOC	Marine Multizone	Marine structures capable of wave damping and corrosion prevention
MITI	Marine Community Polis	Promotion of combined research & development industrial location
MOT	Offshore Man-made Island	Creation of new national area and ocean space
Fishery Agency	Marinovation	Improvement and development of the coastal zone around fisheries

Figure 4 shows a concept of the Marinovation Plan.

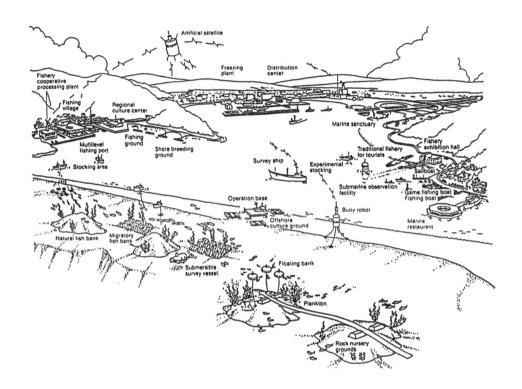

Figure 4.   A concept of the marinovation plan.

THE USE OF THE ENVIRONMENT IN THE DESIGN OF DEEP OCEAN WATER RESOURCE SYSTEMS

J.P. CRAVEN, W.E. BUCHER, and P.K. SULLIVAN
Common Heritage Enterprises
Oceanit Laboratories Inc.
Honolulu, Hawaii
USA

The ocean is the world's largest solar energy collector and absorbs about 10^{13} watts of solar heat. The oceans also are a part of a huge thermodynamic machine that produces ocean currents and waves. When chemical and biological properties of the ocean are also incorporated into this energy machine, the environment becomes a major factor in the design of engineering structures. In general the designer seeks to overcome or minimize environmental forces or effects. He thereby misses opportunities to employ the natural energy and chemistry of the ocean to his advantage. This paper briefly describes three such environmental effects that can be used to great advantage in design. They are the cold, pathogen free water found in the ocean depths, the current and current shear that are generally found in the ocean, and the natural electrochemical deposition of calcium carbonate.

The major energy source of the ocean appears to be Ocean Thermal Energy. It is natural that this tremendous energy source would be considered as an alternative to more conventional energy sources. Recognized as far back as 1882 by D'Arsonval and developed in 1930 by the French engineer Georges Claude (Claude, 1930), the idea of extracting electric power from the heat stored in tropical oceans has enjoyed a resurgence in recent years because of the high cost of fossil fuels. Claude proposed that warm seawater be boiled under a vacuum and that the steam produced be used to turn a turbine and generate electricity. The steam would then be condensed by cold water pumped up from the deep ocean. Claude's concept is called "open-cycle" OTEC and seawater is used as the working fluid. An alternative concept called "closed-cycle" OTEC is based on the use of ammonia or freon as the working fluid; warm and cold seawater are used to evaporate and condense the working fluid.

An OTEC plant must have simultaneous access to warm and cold water with a temperature difference of approximately 20°C. It is therefore logical that most proposed OTEC sites are in the tropical ocean regions where warm water is available near the ocean surface and cold water is available at sufficient depth. Several recent designers proposed floating plants with long cold water pipes (CWP) that reach some 600 to 1,000 m down into the ocean depths to bring up the cold water.

D. A. Ardus and M. A. Champ (eds.), Ocean Resources, Vol. I, 283–290.
© 1990 *Kluwer Academic Publishers. Printed in the Netherlands.*

The mere relocation of the deep ocean water from the deep ocean surface can result in a dramatic change in the local environment. Since the great majority of renewable systems include a phase where the sun's energy is transformed into increased enthalpy of the earth's fluids, the artificially upwelled water will modify this transformation. Thus the rotation of the earth exposes the earth's atmosphere to the sun's rays on a daily basis throughout the year in the equatorial regions and on a semi-annual basis in the polar regions. The net result is a warming in the tropical zones and a cooling in the Artic zones (as a result of radiation into space) which produces circulations in the atmosphere and the ocean of fluid masses having differing temperatures. The existence of land masses makes the circulation non-homogeneous and, where mountains exist, results in vertical movements of masses of air, orographic rain and the collection of water at high altitude in lakes and catchments. As a result each region of the earth is visited by differing temperatures. Nature and man will use these natural heat flows to extract energy in the form of mechanical, electrical and chemical energy. In doing so the amount of energy that can be so extracted is a function of the masses of flow and the efficiency of transfer which is limited by the Carnot Cycle efficiency, $T_1 - T_2/T_{1 ABS}$.

It is generally believed that bio-mass production is simply the result of the presence of water, air and solar photons. This oversimplification neglects the major role played by thermodynamic processes in the transport of water and nutrients from the root zone to the leaf and fruiting areas where photosynthesis takes place. The efficiency of this transport process is also a function of the Carnot efficiency of the thermodynamic fluids that participate in this process and is thus a function of temperature difference that exists in these fluids. The difference in nature and quantity of growth for differing temperature regimes is dramatic.

Two major examples can be given: the first a comparison of the growth associated with "high" islands in the tropical Pacific and the second the change of growth characteristics in temperate zones between the spring, the summer, and the winter.

In the tropical ocean, sea surface temperatures and land surface temperatures are very similar. When trade winds blow on the surface of the sea and the sun strikes the sea surface, evaporative cooling takes place and moisture enters the atmosphere. The resulting temperature profile of the atmosphere is such that vapor condensation in the form of clouds does not take place until an elevation of about 10,000 feet and as a result there is an absence of rain. Thus the surface of the sea and the surface of the low islands are deserts, with the lack of temperature differential being a major contributor to the absence of a beneficial conversion of heat energy to mechanical energy. In the high island, the moist winds are carried to higher cooler elevations where the reduced temperature now produces rain (as much as 600 inches per year at Mount Waialeale in Hawaii). The flows of relatively cold water provide both the fresh water and the temperature difference necessary for plant growth.

In temperate zones in the spring, run-off from winter snows provides the cold ground water that, when combined with the sun's warm rays, results in highly productive growth. As we shall see, it is particularly productive of the high energy product known as sugar. In the fall the ground is warm and the atmosphere cold and again there is a period of high productivity and once more in the form of high energy sugars. During summer the temperature difference and rain results, in general, from the successive passage of high and low pressure weather fronts. The appearance of a cold front will result in precipitation. In the absence of weather patterns, air, water and soil temperatures will become very similar. Growth during this period is largely in the form of low calorie leaf.

Thermodynamic processes also play a role in human vitality and comfort. It is necessary for the human body to maintain an internal temperature of 98.8°F. Optimum life processes suggest an external temperature of about 75°F. Thermodynamic processes in body thus have a Carnot efficiency of about 4%. Colder temperatures require a higher caloric input to maintain body temperature and, if normal oxidation is insufficient for this purpose, shivering will occur as a mechanical means of producing heat. If temperatures are too high, and the body has difficulty in rejecting heat, "heat stroke" is a possible result of vigorous mechanical activity.

It is clear therefore that a productive and comfortable society requires the circulation of fluids having significant temperature differentials in spatial scales which range from the macro-climate, through the micro-climate to the scale of the organism that must use thermodynamic process for its life support functions. When nature does not supply an adequate quantity of process fluid that is sufficiently cool, then the process of chilling or refrigeration is very inefficient. Refrigeration cycles generally consist of the compression of a process gas either by a mechanical compressor or by the addition of heat. The compression process raises the temperature of the gas which now requires cooling by some fluid whose temperature is above that which is desired until the gas liquefies. The liquid at room temperature is now expanded through a nozzle and becomes a cold gas and exchanges its cold with the fluid mass that is to be cooled. This fluid mass (atmosphere or fluid) is then employed in the thermodynamic process that required its existence. This refrigeration process is the reverse of the ocean thermal energy process and is therefore subject to Carnot efficiency in the sense that the energy availability of the cold fluid has been produced with an efficiency that is not greater than the Carnot efficiency associated with the temperature of the cooling fluid in the condenser and the fluid that has been cooled. If this cold fluid is now used in some process that converts thermal energy into some other form of energy, efficiency is further degraded by the Carnot efficiency. Thus, in the growth processes previously described, the use of artificial refrigeration requires a production of energy at an efficiency which is roughly proportional to the square of the Carnot efficiency or, for most operations, an efficiency of less then 1%.

If, on the other hand, a process fluid is available in large quantities that is much colder than is required for the thermodynamic application and that has little or no cost associated with its acquisition and distribution, the combination of energy saving in the generation of cold and energy production in the generation of energy product is proportionate to the reciprocal of the square of the Carnot efficiency.

This happy situation is obtained when deep ocean water is recovered for the primary purpose of oceanic mariculture. Deep ocean water throughout the world has the characteristic that it is cold (approximately 4°C), it is rich in nutrients, and it is biologically pure. The last of these two characteristics makes the fluid of economic value in the production of such high valued warm water crops as grassileria and ogo (commercially significant sea weeds) and a number of commercially significant algae (spirulina, dannaniella). Other biologic products that grow at reduced temperature but well above the temperature of the deep ocean water, such as nori, kelp, and abalone (13°C), do not require the cold associated with the deep ocean. Ocean thermal energy plants will, of course, benefit from the coldest possible water (practically about 6°C) but will warm the cold water about 3°C in the process of generating energy. There is thus available a fluid that may be warmed from 8°C to 13°C without changing its commercial value. Except for transportation and distribution costs this water is essentially economically free.

The Keahole Point Natural Energy Laboratory of Hawaii has been exploring a number of these "economically free" uses of deep ocean water. In their totality they represent the thermostatic regulation of a society. Most dramatic has been the ability to grow high quality spring crops such as strawberries, lettuce, alpine flowers, asparagus, etc.

The technique employed here is to pass the deep ocean cold water through pipes which are embedded in the soil at the root zone. This makes the soil cold and produces fresh cold water in the form of condensate. Thus the optimum temperature for growing such crops as strawberries can be maintained. After the strawberries have fruited the cold water flow is increased, the temperature of the ground is decreased, and the strawberries become dormant. After the appropriate period of dormancy the flow is decreased and the fruiting begins again. Thus strawberry crops can be produced throughout the year independent of the season. Similarly other spring flowers and crops have been produced in the "waterless tropical" environment.

This water may also be employed in a heat exchanger for air conditioning. At the Keahole Laboratory the chilled water air conditioning system has been modified by the addition of a titanium heat exchanger for the exchange of heat between the chill water and the deep ocean water. Calculations by Daniel suggest that cold water flows of approximately 100 gpm will displace approximately 7.5 tons of refrigeration. The bottom line for this installation is empirical and reflected in the cost of operation of the system. For the plant in question the cost of operation with deep cold water is about 25% of the cost of operation of the unmodified plant. Since the air conditioning is a by-product of an energy aquaculture system the capital investment which must be amortized consists only of the piping and heat exchanger associated with the air conditioning plant, and the water cost is simply the pumping cost associated with the additional loop.

Alastair Johnson of Marconi Engineering has also suggested that fresh water can be obtained by the simple extraction of condensate on finned heat exchangers through which the deep ocean water flows. Although this process is highly inefficient as compared with almost any other form of distillation, such water production is also a by-product, and the only element of capital cost for amortization is the heat exchanger and associated plumbing. If so, then this water becomes economically competitive with other desalinization methods.

Thus, deep ocean water as a resource can be used in three modes: in the temperature range from 6°C to 9°C for the generation of electrical power, in the temperature range from 9°C to 14°C as an environmental air conditioner for the simulation of microclimates conducive to growth and physical well being, and in the temperature range from 14°C to about 25°C for acquaculture.

This is but one example of environmental design. Another approach employed by the authors relates to floating Deep Water Ocean Thermal Energy Plants. For example, an OTEC plant that produces hydrogen or ammonia does not need to be attached to a cable and could be allowed to move over a large area of the ocean's surface as long as the motion could be controlled.

To maintain a floating plant within the desired area, a station keeping system is necessary to prevent off-station drift caused by wind, waves and current. An acceptable solution to the station keeping problem would be a dynamic positioning system that has low power requirements but is still able to keep the platform in a desired location or on a desired path.

Craven and Lee (1980) have proposed such a system in which two large lifting surfaces, herein referred to as vanes, similar to large rudders are attached to the cold water pipe (CWP). Where the current has a vertical speed gradient and/or a change in direction with depth, it may be possible to position a pair of properly designed vertical vanes in the current such that the resultant hydrodynamic force has an upstream component. This force will move the platform against the surface current as well as lateral to it. The idea is analogous to what happens when a sailboat tacks upwind with one vane acting as a sail and the other as the keel of the boat.

Thampi (1984) tested Craven's concept with a 40 MW plant configuration. He developed a computer model that calculates steady state plant velocity for various positions of the two vanes relative to the onset current. He found that the plant could have a maximum upstream velocity component that varied from 2.7 to 15.4 cm/sec depending upon the vane angle settings and the drag coefficient of the plant main body.

Butcher (1988) extended Thampi's work and solved the unsteady equations of motion to find the velocity and path of the platform as it accelerates from one steady condition to another. He considered a modified baseline OTEC plant and a wider range of current environments.

Butcher selected a 40 MW spar buoy platform as the dynamic positioning model for his study. He numerically simulated the motions of various spar buoy configured platforms in three current environments - at Kahe Point and Keahole Point in Hawaii and in the Gulf Stream off

Florida. Surface current speeds ranged from less than 0.5 knots to
about 4.5 knots. Simulations were made to determine if the plant could
develop a velocity component that was upstream to the surface current,
then to determine if a plant moving upstream could change tack and
continue to move upstream. These simulations showed that the platforms
achieved maximum upstream velocities of 8.5 cm/sec in the Kahe Point
current and 16.5 cm/sec in the Gulf Stream, and that they could be
maneuvered in these currents.

The main conclusion drawn from this study is that it is possible,
under certain limiting conditions, to dynamically position an OTEC
plant using forces derived from ocean currents. The limiting
conditions are a function of the current environment and the
configuration of the plant. When the conditions are met, the action of
the current on a pair of vertical vanes, which are attached to the CWP,
will produce an upstream force on the plant and a resulting upstream
velocity component.

A third example of environmental design is in the use of
electrically deposited calcium carbonate to make structures. Deep
water is a valuable resource not only because of its nutrient rich,
pathogen free and cold condition, but also because of minerals
contained therein. Through processes of electrodeposition, minerals in
the water column can be recovered to create artificial reefs. This has
been done in surface seawater for several years where the resulting
calcium carbonate aggregate provides a natural habitat for marine
species - closely replicating natural coral reefs.

In addition to artificial reef structures, electrolytically
deposited minerals can be used to create undersea structures,
e.g. protective jackets, and low pressure undersea diver habitats.
Furthermore processes of electrodeposition can be used to create
materials that can be employed in the on-land construction industry,
e.g. building blocks, poles, etc. This is particularly important in
land-resource poor societies such as is found in the developing
Pacific.

The ability to control the rate of deposition is a matter of
controlling the equilibrium kinetics of the deposition reaction:

$$CaCO_3 == Ca_2^+ + CO_3^{-2}$$

By electrolyzing water, the pH on the surface of the cathode
increases, causing the precipitation of calcium carbonate. Typical
seawater contains 400 mg/kg of calcium; however, deep seawater
recovered from OTEC depths of approximately 2,000 feet contains more
calcium per unit weight, $CaCO_3$ precipitates in warm subtropical
environments, i.e. beaches in Hawaii. However, dissolution occurs at
the depths due to pH and temperature considerations. Therefore, we
expect more calcium per unit weight from deep seawater.

Cold water pumped from the depths has interesting characteristics
that may influence the kinetics of deposition, i.e. low pH, temperature
and total organic carbon, as well as high carbon dioxide. Through
various processes including photosynthesis and degassification the pH
can be increased so that precipitation is enhanced. However, this

analysis rapidly leads to the need for empirical testing so that rate
determining steps and significant chemical species can be determined.

Calcium carbonate deposits have been found to have an average
compressive strength of up to 4300 psi. For comparison, concrete will
fail at about 3500 psi (probable 28th day strength, normal Portland
cement Type 1, 15% less than recommended by ACI Joint Committee)
(Hilbertz, 1978).

Mitsui Shipbuilding in Tokyo, Japan, has developed techniques to
deposit calcium carbonate. Results from their research and testing
indicate that the main factors affecting the deposition are electric
current density, water current, water temperature and the surface
conditions of the materials of the cathode. The weight of the
deposited minerals is nearly proportional to period and intensity of
electric current. The inclusion of magnesium in the $CaCO_3$ matrix is
dependent on current density. At lower currents there is more of $CaCO_3$
and the hardness and the bending strength are greater. At higher
current densities the amount of $Mg(OH)_2$ increases, where the hardness
and bending strength decrease.

Table 1 compares the properties of typical electrolytically
deposited materials found by Mitsui. There testing has shown that when
the amount of $CaCO_3$ is approximately 50% of the aggregate deposit the
strength of the deposition is close to that of concrete. The thermal
conductivity of the deposition is 1/30 to 1/40 of that found for steel.

TABLE 1. Characteristics of Electrolytically Deposited Aggregates

		Electrolytically Deposited Aggregate	Concrete
Compressive strength	(kg/cm³)	60-200	200-500
Bending strength	(kg/cm³)	50-70	30-60
Heat conductivity	(kcal/mh-°C)	1.18	1.3
Young's ratio	(kg/cm³)	2×10^5	3×10^5
Specific gravity		2.5	2.8

REFERENCES

Bucher, W. (1988) 'Dynamic positioning of a floating OTEC plant
    using ocean environmental forces`, Ph.D. Dissertation, Univ. of
    Hawaii.

Claude, G. (1930) 'Power from the tropical seas`, Mechanical
    Engineering, Vol. 52, No. 12, pp.1039-1044.

Craven, J.P. and Lee, T.D. (1980) 'Stable platforms for ocean
    energy applications`, Proc. 16th Marine Technology Society
    Annual Conf., Washington DC.

Thampi, S. (1984) 'Dynamic positioning of an OTEC plant using
    environmental forces`, M.Sc. Thesis, Univ. of Hawaii.

PART VII

Environmental Assessment

OBTAINING PAST TRENDS IN MARINE ENVIRONMENTAL CONDITIONS WITH
CONTEMPORARY DATA

CHENG-TUNG A. CHEN
Institute of Marine Geology
National Sun Yat-Sen University
Kaohsiung, Taiwan
REPUBLIC OF CHINA

ABSTRACT. Two major approaches have traditionally been used to obtain
secular variations of pollutant concentrations in the past. The direct
approach is to compare current data with data collected at different
times. This approach, however, is often unreliable because few earlier
data exist. Furthermore, those that are available are sometimes of
poor precision or with doubtful accuracy. As a result, comparing these
unreliable data with current reliable data does not always give a
meaningful trend. The indirect approach is to obtain samples of
various ages. By analyzing and dating these samples, one obtains the
past trend. Analyzing and dating seawater and sediment samples are
notable examples in oceanography and geology. Another indirect
approach is to analyze pollutant concentrations in banded corals. A
case study for trace metals and radionuclides is presented.

1. INTRODUCTION

With the establishment of the 200-mile Exclusive Economic Zone (EEZ),
exploitation of marine resources within EEZ is bound to increase. And
with increased activities, there comes increased pollution and the
potential for conflicts regarding the use of EEZ. From the scientific
point of view, it is important to know how these human activities
change the marine environment. From the practical point of view, it is
also desirable, or necessary, to know whether the activity of a certain
industry (or even better, a specific company) has added significant
amounts of pollutants into the marine environment. It is therefore,
important to establish a baseline for pollutant concentrations. Even
if the baseline has already been altered by previous activities, we
must still find a way to obtain the past trends of pollutant
concentrations.

Generally speaking, there are two major approaches to obtain the
past trends in the marine environment. The first approach compares
present data with data obtained from previous years. The difference
represents the variation of pollutant concentrations with time. The
method is straight forward but has a major drawback when applied to
chemicals of trace amount, because early data are scarce and frequently

293

D. A. Ardus and M. A. Champ (eds.), Ocean Resources, Vol. I, 293–305.

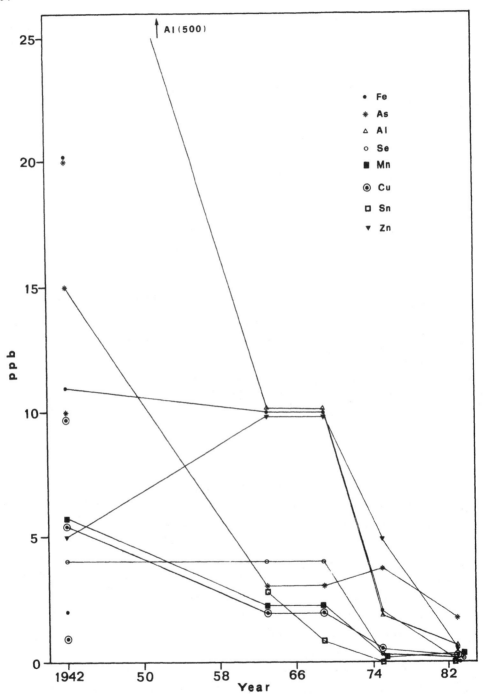

Figure 1. Concentrations of selected trace metals in seawater reported
in different years (data from Goldberg, 1963; Horne, 1969; Brewer,
1975; Bruland, 1983).

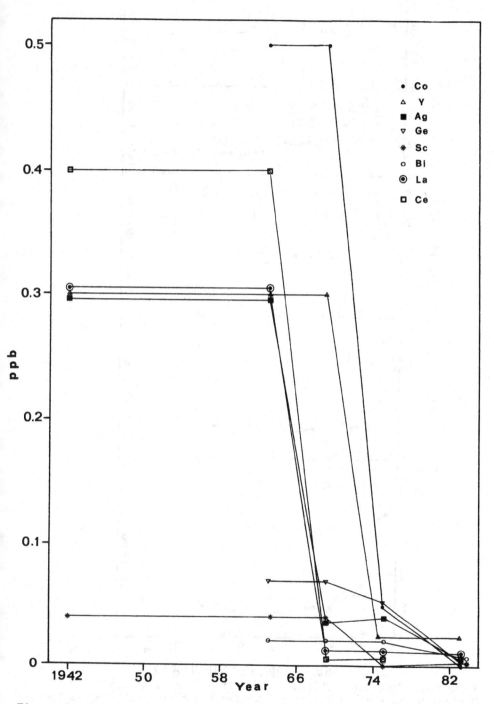

Figure 1.   Continued.

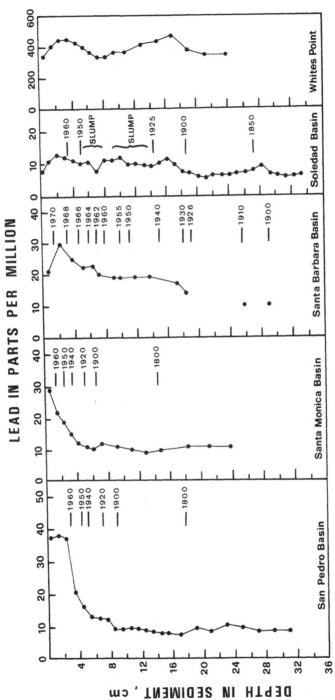

Figure 2. Vertical profiles of lead in sediments at San Pedro, Santa
Monica, Santa Barbara Basins and Whites Point off southern California,
and at Soledad Basin off Baja California (taken from Goldberg, 1975).

unreliable. As a result, direct comparison of the present data with
early data which have large uncertainties does not always give a
meaningful trend.

Take heavy metals as an example, early samples were often
contaminated during sampling, making the measured values artificially
high. I have plotted seawater trace metal concentrations taken from
text books published in different years (Fig. 1). The aluminium
concentration reported in 1983 was only a quarter of the 1975 value.
In other words, it was believed in 1983 that the aluminium
concentration reported only eight years ago was a factor of four too
high. Many other trace metals were also found to be much higher in
1975 than in 1983, and even higher in 1969 or 1963. If one took these
results at face value, then one would arrive at the conclusion that
trace metal concentrations in the marine environment are decreasing,
which is almost certainly an erroneous conclusion.

The second major approach to obtain the past trends of pollutant
concentrations in the marine environment is to use sediments. The
vertical profiles of lead in sediments collected near southern
California are plotted in Fig. 2 as an example. In general, the deeper
the sediment, the older it is. It is obvious that the lead content is
the highest near surface in the San Pedro, Santa Monica and Santa
Barbara Basins. The values approach a constant, representing a
baseline, for deep sediments deposited prior to 1900. The Soledad
Basin off Baja California is away from major pollution areas and the
lead concentration is relatively low throughout the sediment column.
On the other hand, the Whites Point, which receives the Los Angeles
area sewage outflow, has very high lead concentration in the sediments,
as indicated clearly in the figure.

Sediments, however, only record what settles down, either as
particles or as material absorbed on the surface. Sediments are also
frequently re-worked or undergo diagenesis, thus disturbing the
original signal. Further, dating sediments is not a trivial matter and
it is difficult to obtain annual bands. Two other approaches have
recently been used to obtain the past trends of pollutant
concentration. The first approach is used to trace back pollutant
concentrations in the water column, and the second provides accurate
dates.

2.   ANTHROPOGENIC SIGNAL IN SEAWATER

Deep and Bottom Waters are mainly formed in the polar regions. Once
they sink below the air-sea interface, it may take hundreds of years
before the water returns to the surface and contacts the atmosphere
again. Consequently, subsurface waters of various ages record certain
pollutant concentrations in the atmosphere when the water were last in
contact with the atmosphere. Thus tracing waters of various ages
provides a trend of pollutant variations especially if the pollutant
remains in the dissolved phase and moves with water.

Figure 3 provides an example of the distribution of bomb-produced
tritium in the Atlantic Ocean (Ostlund et al, 1976). Tritium
penetrates all the way to the seafloor in the northern North Atlantic.

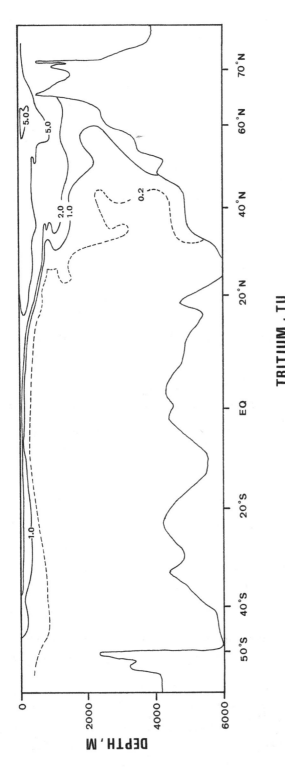

Figure 3. Cross-section of tritium in the Atlantic Ocean (taken from Ostland et al, 1976).

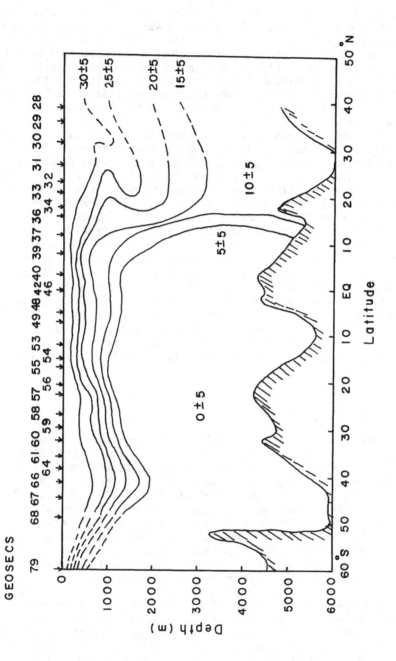

Figure 4.   Cross-section of anthropogenic CO₂ in the Atlantic Ocean (taken from Chen and Drake, 1986).

Further, there is a tritium minimum (older water) at an intermediate
depth where remnant Antarctic Intermediate Water (AAIW) can be found.
Tritium moves freely with water masses and does not undergo chemical
reactions except by radioactive decay. Some chemicals, such as C-13
and CO_2, however, are altered by the production and decomposition of
organic material and calcium carbonate. (Chen and Millero, 1979;
Chen, 1982; Kroopnick, 1985). Once these alterations are subtracted
from the observed values, the contribution of the anthropogenic
activities appears.

Figure 4 shows a cross-section of excess, anthropogenic CO_2 in
the Atlantic Ocean (Chen and Drake, 1986). Deep waters south of 10°S
do not contain any excess CO_2 because these waters were formed prior to
industrialization and has not contacted fossil-fuel CO_2. Near-surface
waters and deep waters in the North Atlantic Ocean, however, were
formed more recently, thus contain excess CO_2. The highest amount of
excess CO_2 is found in the surface waters now in contact with the
atmosphere. An excess CO_2 minimum is found at the tritium minimum.

When fossil fuels are burned, the CO_2 released contains more C-12
relative to C-13 as compared to the atmosphere. As a result, the C-
13C-12 ratio in the atmosphere has gradually decreased, and this signal
can be traced in the oceans. Fig. 5 shows the preformed C-13
concentration for the Antarctic Intermediate Water in the Atlantic,
Indian and Pacific Oceans (Chen and Chen, 1988). The AAIW is formed
near the Subtropical Front and then spreads northward. Fig. 5
indicates that when older AAIW waters near the equator were formed,
they contained more C-13 than more recently formed waters further
south.

3.   ANTHROPOGENIC SIGNAL IN CORALS

Several orders of corals form annual growth bands which can be used to
study the environmental perturbations in both contemporary and paleo-
communities. For example, patterns of annual banding has been used to
assess the impact of long-term changes involving sedimentation,
temperature and nutrients (Knutson et al, 1972; Hudson et al, 1976;
Dodge et al, 1984). Several investigators have also attempted
recently, to varying degrees of success, to use banded coral skeletons
to document the chronology of coastal pollution (Dodge and Gilbert,
1984; Chen et al, 1986, 1988; Shen and Boyle, 1987, 1988; Shen et al,
1987). For instance, annual variations of several metals have been
detected for samples collected in Southern Taiwan.

Figure 6 shows the variation of zinc and cadmium, both used
heavily in industry (Chen et al, 1986; Cheng, 1988). Zinc shows the
clearest increase with time and the values are higher for samples
collected at Hsiao-Liu-Chiu near the heavily polluted industrial city
of Kaohsiung. Local fishermen have complained that the dumping of
slags from the China Steel Corp. off Kaohsiung starting in 1985 has
increased the trace metal content, especially zinc, in seawater. Our
data, however, indicate that the zinc concentration has started to
increase at least 20 years ago. China Steel Corp., therefore, could
not have been the culprit. Samples collected at the Third Nuclear

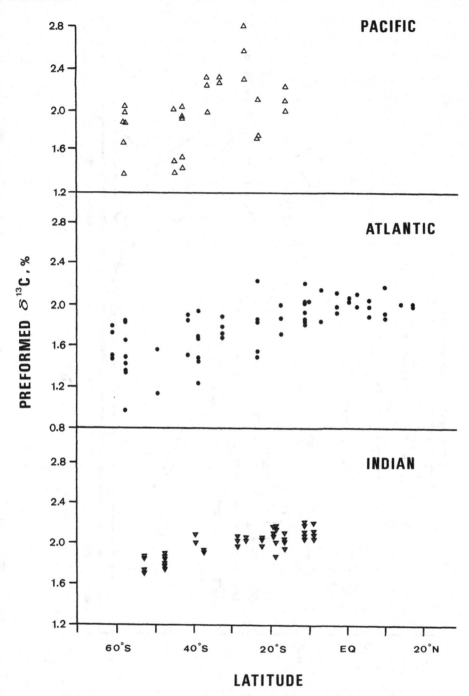

Figure 5.   Preformed δC-13 concentrations for the Antarctic
Intermediate Water.

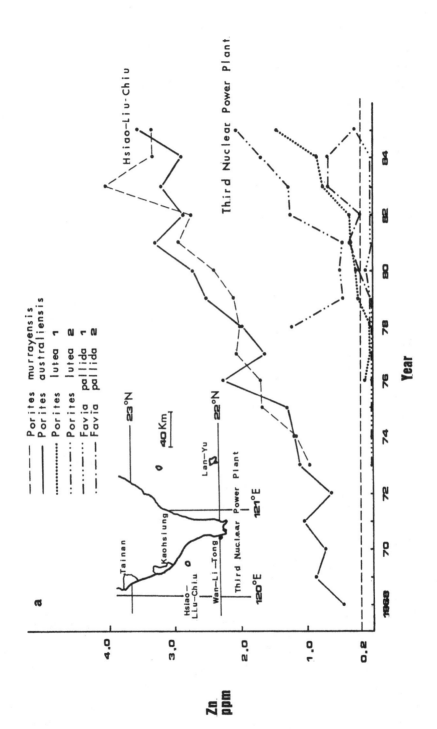

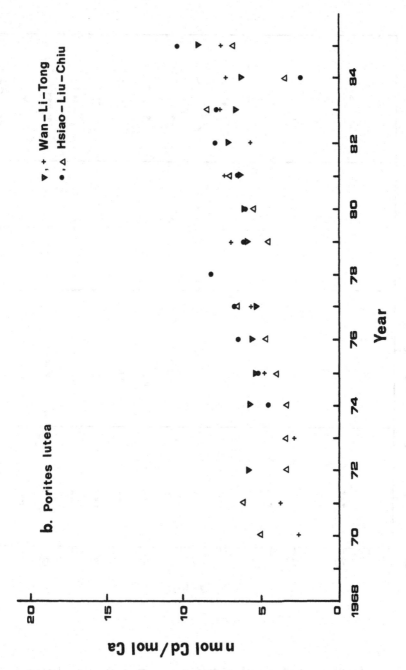

Figure 6.   Annual variation of (a) zinc and (b) cadmium in corals
collected in Southern Taiwan (taken from (a) Chen et al, 1986; and
(b) Cheng, 1988)

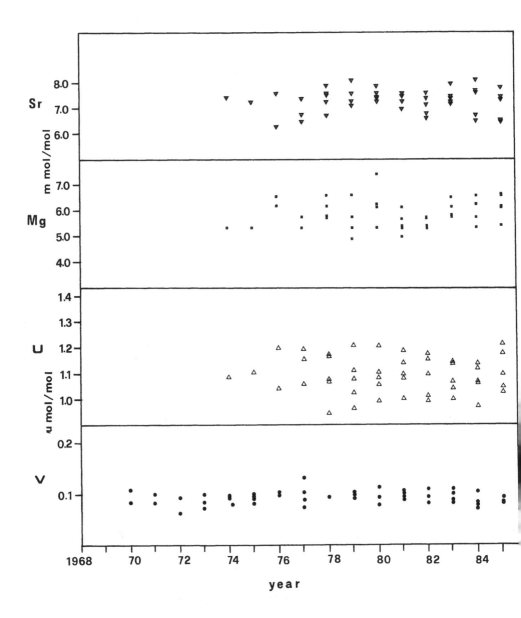

Figure 7. Annual variation of strontium, magnesium, uranium and
vanadium in corals collected in Southern Taiwan (taken from Chen,
unpublished results, 1986; and Cheng, 1988).

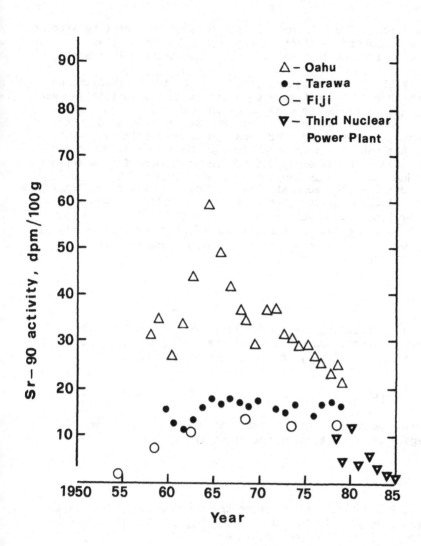

Figure 8.   Annual variation of Sr-90 in corals (taken from Chen et al, 1988).

Power Plant contain lower amount of zinc because the sampling site is
far from any industrial activities except for the power plant.
Cadmium shows a less clear trend but the concentration is also
increasing.

Figure 7 shows the concentrations of strontium, magnesium, uranium
and vanadium that we did not expect to detect annual variations because
the anthropogenic inputs are swamped by the much higher natural
background. Indeed we did not see any annual variations for these
metals. Sr-90, on the other hand, was produced by a large amount
relative to the natural environment during the atmospheric nuclear bomb
tests which peaked in the early 1960s. The concentrations started to
decrease since then and our samples collected near the Third Nuclear
Power Plant confirm the decreasing trend reported elsewhere (Toggweiler
and Trumbore, 1985). Since other artificial nuclides commonly
discharged by a nuclear power plant, such as Mn-54, Co-60, Zn-65,
Sb-125 and I-131, are undetectable in our samples, we believe that the
Sr-90 comes mainly from the atmospheric fallout and the level is
decreasing (Fig. 8).

4. CONCLUSION

Certain anthropogenic signals can be detected in seawater and in
corals. These signals can be used to establish past base-lines and to
reveal the pollution history of the marine environment.

5. ACKNOWLEDGEMENT

I acknowledge the support of the National Science Council and the
Nuclear Regulatory Committee of the Republic of China.

6. REFERENCES

Brewer, P.G. (1975) 'Minor elements in sea water', in Chemical
     Oceanography, V.1, 2nd ed., J.P. Riley and G. Skirrow, eds.,
     Academic Press, London, pp.415-496.

Bruland, K.W. (1983) 'Trace elements in sea-water', in Chemical
     Oceanography, V.8, 2nd ed., J.P. Riley and R. Chester eds.,
     Academic Press, London, pp.157-220.

Chen, C.T. (1982) 'On the distribution of anthropogenic CO_2 in the
     Atlantic and Southern Oceans', Deep-Sea Research 29, pp.563-580.

Chen, C.T. (1987) 'Obtaining past trends in pollutant
     concentrations with contemporary data', Bulletin of Marine
     Science and Technology, 2, pp.38-48 (in Chinese).

Chen, C.T.A. and Millero, F.J. (1979) 'Gradual increase of oceanic
     CO_2', Nature 277, pp.305-307.

Chen, C.T. and Drake, E. (1986) 'Carbon dioxide increase in the atmosphere and oceans and possible effects on climate', Ann. Rev. Earth Planet, Sci. 14, pp.201-235.

Chen, C.T., Fang, L.S., Chang, K.H., Chou, E.C. and Hwang, J.C. (1986) 'Annual variation of heavy metals and radio-isotopes in corals in recent years', Proceedings, symposium on Marine Sciences, Kaohsiung, Taiwan, Republic of China, August, pp.141-147 (in Chinese).

Chen, C.T.A., Chang, K.H. and Fang, L.S. (1988) 'Recent changes of trace metals and radionuclides in banded corals in southern Taiwan', Environmental Protection, in press.

Chen, D.W. and Chen, C.T.A. (1988) 'The anthropogenic CO_2 signal in the Indian Ocean', Environmental Protection, submitted.

Cheng, K.L. (1988) 'Method of trace metal pre-concentration and analysis in seawater and corals', thesis, Sun Yat-Sen Univ., Kaohsiung, Taiwan, 97pp.

Dodge, R.E. and Gilbert, T.R. (1984) 'Chronology of lead pollution contained in banded coral skeletons', Mar. Biol., 82, pp.9-13.

Dodge, R.E., Jickells, T.D., Knap, A.H. and Bak, R.P.M. (1984) 'Reef-building coral skeletons as chemical pollution (phosphorus) indicators', Mar. Pollut. Bull., 15, pp.178-187.

Goldberg, E.D. (1963) 'The oceans as a chemical system', in The Sea, Vol. 2, M.N. Hill ed., Interscience, New York, pp.3-25.

Goldberg, E.D. (1975) 'Marine pollution', in Chemical Oceanography, J.P. Riley and G. Skirrow, eds., Academic Press, London, pp.39-89.

Horne, R.A. (1969) 'Marine Chemistry', Wiley-Interscience, New York, 568 pp.

Hudson, J.H., Shinn, E.A., Halley, R.B. and Lidz, B. (1976) 'Sclerochronology, a tool for interpreting past environments', Geology, 4, pp.361-364.

Knutson, D.W., Buddemeier, R.W. and Smith, S.V. (1972) 'Coral chronometers: seasonal growth bands in reef corals', Science., 177, pp.270-272.

Kroopnick, P.M. (1985) 'The distribution of ^{13}C and CO_2 in the world oceans', Deep-Sea Res. 32, pp.57-84.

Ostlund, H.G., Dorsev, H.G. and Brescher, R. (1976) 'Tritium Laboratory Report No. 5, Univ. Miami, 192 pp.

Shen, G.T. and Boyle, E.A. (1987) 'Lead in corals: reconstruction
    of historical industrial fluxes to the surface ocean', Earth
    Planet Sci. Lett. 82, pp.289-304.

Shen, G.T. and Boyle, E.A. (1988) 'Determination of lead, cadmium
    and other trace metals in annually-banded corals', Chem. Geol.
    67, pp.47-62.

Shen, G.T., Boyle, E.A. and Lea, D.W. (1987) 'Cadmium in corals as
    a tracer of historical upwelling and industrial fallout',
    Nature, 328, pp.794-796.

Toggweiler, J.R. and Trumbore, S. (1985) 'Bomb-test Sr-90 in
    Pacific and Indian ocean surface water as recorded by banded
    corals', Earth Planet Sci. Lett. 74, pp.306-314.

POTENTIAL USE OF CIRCULATION/POLLUTANT TRANSPORT MODELS FOR IMPACT
ASSESSMENT ON THE U.S. EEZ

ROBERT P. LaBELLE
Minerals Management Service
U.S. Department of the Interior
U.S.A.

ABSTRACT.   This paper presents an overview of how applied oceanography
may contribute to quantitative efforts in estimating the environmental
hazards of exploration and development of nonenergy mineral resources
in the Exclusive Economic Zone (EEZ) of the United States. The
oceanographic research program which currently supports oil spill
trajectory modelling on the U.S. Outer Continental Shelf is outlined
with emphasis on the application of study results to U.S. EEZ areas.
Several three-dimensional circulation models presently used to
represent conditions offshore the conterminous States and Alaska are
described. The spatial and temporal domains of these models are
compared to U.S. EEZ boundaries in order to determine areas of overlap
and therefore potential use.
        Also discussed is the use of simulation modelling to represent the
locations of biologic and economic resources, as well as potential
pollutant transport in the EEZ. Risk estimates developed from such
analyses may contribute to leasing decisions, such as whether to impose
stipulations or to delete certain areas from development. In addition,
quantitative assessments of risks are needed to describe and analyse
the effects of alternative leasing proposals.

1.   INTRODUCTION

Exploration and development of marine minerals in the Exclusive
Economic Zone (EEZ) should represent a reasonable balance between the
national interest and the potential for environmental damage or adverse
impact to coastal and marine resources. Currently in the United
States, the Minerals Management Service (MMS) is determining policy for
exploration, development, and recovery of energy and nonenergy minerals
from the U.S. EEZ. Leasing proposals are developed in cooperation with
the adjacent coastal States.
        To ensure protection of the environment, a case-by-case approach
is planned that incorporates a series of environmental reviews with
opportunities for public involvement and environmental studies, as
applicable. This paper will address the use of existing environmental

309

D. A. Ardus and M. A. Champ (eds.), Ocean Resources, Vol. I, 309–319.
© 1990 Kluwer Academic Publishers. Printed in the Netherlands.

information and models to assist in evaluating risks from activities associated with the extraction of marine minerals.

Applied oceanography has greatly improved the capability of analysts to quantitatively assess potential impacts to the marine environment from offshore oil and gas activity. One of the ways that the MMS complies with its mandate of protection of the environment concomitant with minerals resource development is by sponsoring the Outer Continental Shelf (OCS) Environmental Studies Program. This is the largest single-agency, mission-oriented oceanographic research program in the Federal Government. Since its inception in 1973, the program has funded over $100 million on physical oceanography studies of the OCS environment.

A primary use of the data generated by these studies has been to develop and enhance information used in oil spill trajectory modelling. Model results are used to analyse risks to areas of ecological or economic concern and play an important part in leasing decisions, such as to impose stipulations or to delete individual tracts. Much of the information generated by these studies may also be useful for risk assessment of marine mining activities in the EEZ. For example, existing ocean circulation models provide information on subsurface transport that could be used to assess risk to bottom habitats from dredging operations.

In addition to modelling the potential movement of pollutants, it is necessary to develop a representation of the locations of biologic and economic resources at risk in a given study area. Leasing activities in the EEZ may be preceded by impact analysis on such resources as infauna, epifauna, and various habitat types within the lease region; species with planktonic larval stages; marine mammals; and commercial and subsistence fisheries which may be affected by destruction of bottom habitat or reduction of food organisms. This paper suggests that existing procedures now in use for assessing risks from oil and gas leasing to such resources are easily adapted to analysis of potential impacts of marine mining on the EEZ.

2.   CIRCULATION MODELS CURRENTLY AVAILABLE

While MMS and other agencies have funded direct field measurements of offshore current and hydrographic parameters, adequate coverage of temporal and spatial variability is not feasible or cost effective. An alternative for establishing accurate simulations of transport and velocity patterns within areas of interest is a computer-based numerical model of circulation, with appropriate physical considerations and governing equations of motion.

Circulation models in use and planned for use in oil spill trajectory analysis work at MMS have been outlined in LaBelle, 1986, and LaBelle and Anderson, 1985. While predominantly directed toward deriving surface currents for oil spill movement, these studies have also resulted in subsurface and bottom current representations. Such products are presently available for use in modelling of subsurface plumes or other pollutant transport that might result from marine mining operations.

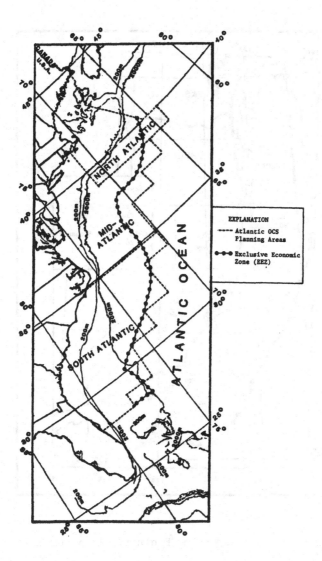

Figure 1. Atlantic boundaries of the Exclusive Economic Zone and the
OCS Planning Areas.

As shown in Figures 1 and 2, boundaries of the Atlantic and
Pacific OCS Planning Areas overlap large portions of the U.S. EEZ.
Calculated seasonal currents in these regions are available to MMS from
the General Circulation Model (GCM) of Dynalysis of Princeton (Blumberg
and Mellor, 1983). The model, run in a diagnostic mode, uses the full
equations of motion, with a two-variable turbulence model to determine
vertical mixing and temperature and salinity fields from seasonal
climatology.

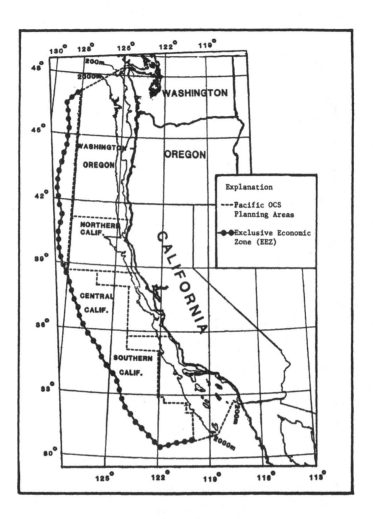

Figure 2. Pacific boundaries of the Exclusive Economic Zone and the OCS
Planning Areas.

The Dynalysis model employs a variable grid size orthogonal
curvilinear coordinate system which facilitates increased resolution
near the coast. Figure 3 shows an example of current vectors from this
model at a depth of 250 m off the Atlantic coast. Figure 4 shows
spring mean currents derived from the diagnostic GCM at 200 m depth
offshore California. At this depth, the Ekman wind-driven component is
strongly attenuated, resulting in a clear picture of the barotropic and
density-driven current patterns. The model also portrays relatively
strong currents at 1,000 m depth along the continental slope that
appear to be barotropic in nature, as they are also evident near the
bottom of the water column (Blumberg et al, 1984).

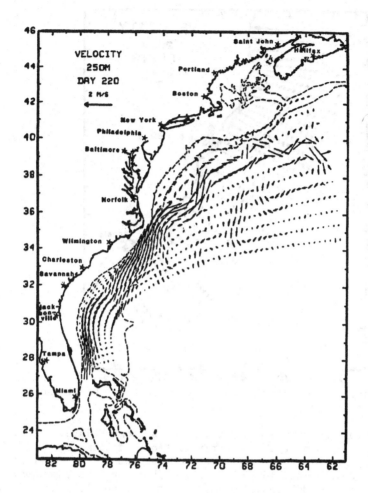

Figure 3. Example of current vectors at 250 m from a benchmark
calculation (Dynalysis of Princeton, May 1988 Progress Report, Ocean
Circulation Model for U.S. Atlantic Coast and Florida Straits).

Beyond the curvilinear grid domain, for example, off the coasts of
Oregon and Washington, currents are obtained using the Dynalysis
Characteristic Tracing Model (CTM) (Kantha et al, 1986). This model,
based on the conservation of planetary potential vorticity, is driven
with the same seasonal climatological hydrography as the GCM, but the
resolution is a uniform half-degree grid. Given the extremely large
regions of oceans that are modelled, the CTM makes use of hydrographic
measurements which have relatively large spatial and temporal coverage
and are readily obtained from historical data archives. While the CTM
may be used to provide a large-scale perspective on seasonal currents,
its primary function is to provide the open ocean boundary conditions
for the more sophisticated, three-dimensional, time-dependent GCM.

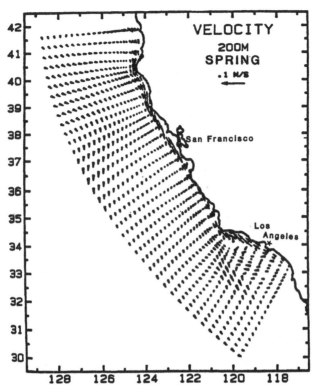

Figure 4. Spring mean currents at 200 m depth from the diagnostic GCM.
(Dynalysis of Princeton).

In the Gulf of Mexico region, MMS has supported Jaycor, Inc. in a
4-year program in numerical ocean circulation modelling to
progressively upgrade an existing model developed by the Naval Ocean
Research and Development Activity (NORDA) (Hurlburt and Thompson,
1980). The final model should have a horizontal resolution of about
10 km and a vertical resolution approaching 1 to 10 m in the mixed
layer, 10 m at the thermocline, and 100 m in deep water.

Two examples of the circulation calculated by the NORDA model are
shown in Figure 5. The Layer 1 currents, shown for a specific day from
a long simulation, display the characteristic Loop Current and its
associated eddies and meanders. Of more interest to this paper,
however, is the representation of the Layer 2 currents, as shown in
the lower portion of the figure. These results cover the entire Gulf,
represent deep water flow, and will be calculated on at least a
seasonal basis. Results from this model should be available to MMS in
early 1989.

MMS has used the Coastal Sea Pollutant Transport Model developed
by Applied Science Associates Inc., for simulating oil spill
trajectories in OCS lease areas offshore Alaska. This three-
dimensional, numerical hydrodynamic model computes sea elevation and
currents due to wind, tide and density forcing. A shallow water wave
forecast model provides information on nonlinear induced wave-current

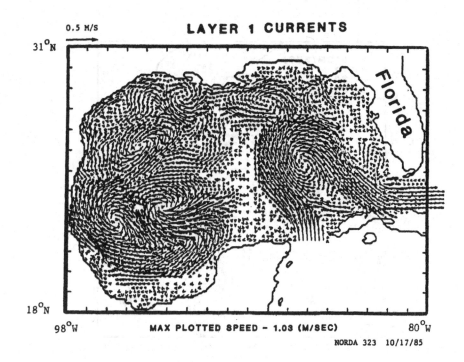

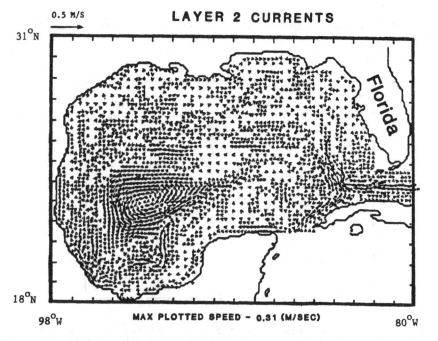

Figure 5. Examples of Layer 1 and Layer 2 currents from the
NORDA/JAYCOR model, representing near-surface and deep-water flow.

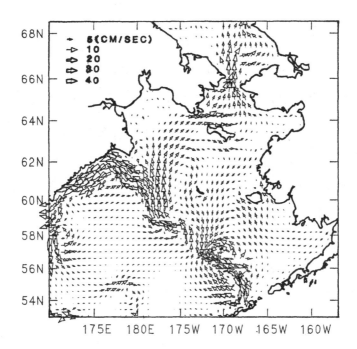

Figure 6. Example of summer density-induced flow at 15 m depth in the
Bering Sea (Applied Science Associates, Inc.).

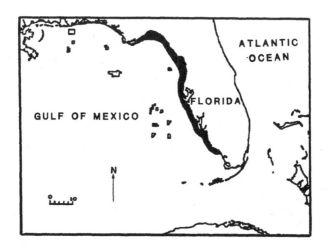

Figure 7. Map showing sample resource in the Eastern Gulf of Mexico.
Shaded area indicates blue crab migration route. Also shown are
examples of potential oil and gas lease tracts.

bottom stress to the hydrodynamic model. The model is capable of
explicitly resolving sub-surface currents, as described in Anderson and
Spaulding, 1986. Figure 6 shows an example of summer density-induced
flow at 15 m depth in the Bering Sea.

Also available to MMS is the Mud Discharge Model (Brandsma and
Divoky, 1976), funded under the Offshore Operators Committee and
developed under the direction of the Exxon Production Research Company.
This model may be used to predict the fate of drilling mud discharges
in the ocean or estuarine environments. Emphasis is on the fate of a
wide range of particle-size mud solids discharged at various rates into
an environment of variable currents, depths, and density
stratification. Plume clouds are tracked until they settle on the
bottom. By using appropriate input characteristics of the pollutant to
be tracked, including modifications to the nature of the release event,
this approach might prove useful in marine mining impact assessments.
For example, a modified version of this model may be able to provide
estimates of the thickness and extent of dredge-generated sediments on
bottom areas.

3.   MODELLING RESOURCES AT RISK

Results from the above models are capable of being processed through
the Oil Spill Risk Analysis (OSRA) model of MMS (Smith et al, 1982).
This model, using an efficient co-ordinate transformation procedure to
incorporate trajectory information provided by external circulation
models, traces hypothetical oil spills through a matrix of grid cells.
The grid matrix also stores the locations of biological and
recreational or economic resources that may be vulnerable to oil, the
representation of the coastline in the study area, and the potential
locations of offshore lease sites. Given that different trajectory
algorithms are necessary to model the many OCS lease areas under study,
the use of a common tracking model (OSRA) enhances consistency in the
MMS's overall approach to risk assessment.

In like manner, the OSRA may easily be modified to represent risks
to two-dimensional resources (such as bottom habitats or the coastline)
from mining site hazards in the EEZ. This approach allows for
specific, state-of-the-art trajectory models to simulate movement of
pollutants through a model grid which contains the locations of
resources at risk. The contacts to each resource would be recorded,
allowing for the calculation of the statistics of potential impact. In
the case of three-dimensional resources, the OSRA model could be
modified to incorporate a series of grid levels to represent resources
located throughout the water column. Of course, this would dictate
additional computing and storage requirements, but is feasible,
nonetheless.

The OSRA employs a Cartesian coordinate system superimposed over a
base map of the area under study for a given analysis. All stored data
are referenced to this grid system, which is used for all internal
calculations. The base map must be large enough so that all coastal
and at-sea resources likely to be affected by the proposed lease action
are included. This involves a tradeoff, with grid resolution

decreasing as the area modelled is expanded. Typical resolution is in the order of 2 to 4 km per grid cell.

Spatial data, including the presence of land or resources, are digitized from standard Mercator or UTM projections and stored in matrices of grid cells for rapid retrieval and access. The model stores indicators of the locations of up to 32 representations of such resources as live bottom areas, endangered species habitats, and fisheries. Figure 7 shows an example of these representations, plotted along with sample locations of potential offshore oil and gas lease sites. Since many biological populations display seasonal migration, the model accounts for monthly presence or absence of the resource in the study area. For example, spawning areas may only be vulnerable to contact during certain months of the year. Therefore, contacts with trajectories would only be counted for those months.

4.   IMPACT ANALYSIS

The calculated statistics of contact from trajectories modelled as described above represent conditional probabilities in an impact assessment. The "condition" is that a release of a pollutant has in fact occurred, and given that occurrence, the estimate is of where the pollutant is likely to go and what resources are likely to be contacted over a given time period. The chance of the release of pollutant is quite separate from such an analysis and should be considered in combination with the conditional probabilities, for a thorough impact analysis.

In MMS oil spill modelling, conditional probabilities are reported for each lease site for all resources. This information is further broken down into contact probabilities for the specific seasons modelled in the trajectory analysis. These conditional estimates should not be considered "worst case" values, for purposes of impact assessment. The probabilities are calculated from trajectories that represent climatology or representative conditions in the study area over a long period of time. In this way, analysts may use the conditional probabilities as a model of what is most likely to happen in general, if a pollutant is released in a given area.

Estimates of risk of contact to resources may be used to analyse and compare the potential effects of alternative groups of proposed lease tracts. This information may be helpful in determining whether to impose stipulations on operations in a sensitive area or whether to delete the area from leasing altogether. Whether used on the OCS or the EEZ, the method allows for characterization of those lease areas most likely to result in contacts to nearby environmental resources.

5.   ACKNOWLEDGEMENT

The contribution of Eileen M. Lear to the preparation and proofing of this manuscript is appreciated by the author.

6.   REFERENCES

Anderson, E. and Spaulding, M.L. (1986) 'Oil Dispersion from
    Selected Spill Sites in Shumagin Basin`, NOAA Outer Continental
    Shelf Environmental Assessment Program (OCSEAP) Report.

Blumberg, A.F. and Mellor, G.L. (1983) 'Diagnostic and prognostic
    numerical circulation studies of the South Atlantic Bight`,
    J. Geophys. Res., 88, pp.4579-4592.

Blumberg, A.F., Kantha, L.H., Herring, H.J. and Mellor, G.L.
    (1984) 'California Shelf Physical Oceanography Circulation
    Model`, Final Report, Report Number 88.2, Dynalysis of
    Princeton, Princeton, New Jersey.

Brandsma, M. and Divoky, D. (1976) 'Development of models for
    prediction of short term fate of dredged material discharged in
    the estuarine environment`, Contract Report D-76-5, U.S. Army
    Waterways Experiment Station (available from Natl. Technical
    Information Service, Springfield, Va.).

Hurlburt, H.E. and Thompson, J.D. (1980) 'A Numerical Study of
    Loop Current Intrusions and Eddy Shedding`, J. Phys.
    Oceanography, 10, pp.1611-1651.

Kantha, L.H., Herring, H.J. and Mellor, G.L. (1986) 'South
    Atlantic OCS Circulation Model - Phase III`, Final Report,
    Report Number 91, Dynalysis of Princeton, Princeton, New Jersey.

LaBelle, R.P. (1986) 'Use of Applied Oceanography in Stochastic
    Modelling of Oil Spills on the Outer Continental Shelf`, IEEE
    OCEANS '86 Conference Proceedings, September 23-25, 1986.

LaBelle, R.P. and Anderson, C.M. (1985) 'The Application of
    Oceanography to Oil-Spill Modeling for the Outer Continental
    Shelf Oil and Gas Leasing Program`, Marine Technology Society
    Journal, Vol. 19(2), pp.19-26.

Smith, R.A., Slack, J.R., Wyant, T. and Lanfear, K.J. (1982) 'The
    Oil Spill Risk Analysis Model of the U.S. Geological Survey`,
    Geological Survey Professional Paper 1227, 40pp.

ENVIRONMENTAL ASSESSMENT FOR EXCLUSIVE ECONOMIC ZONE MINERAL
DEVELOPMENT ACTIVITIES; THE LESSONS LEARNED FROM OFFSHORE OIL AND GAS
DEVELOPMENT

JEFFREY P. ZIPPIN
Department of the Interior
Minerals Management Service
Reston, Virginia
U.S.A.

1. INTRODUCTION

The 1983 Proclamation on the Exclusive Economic Zone (EEZ) placed
approximately 4 billion acres of the seabed adjacent to the United
States (U.S.) under U.S. jurisdiction and control. Since that time,
and even prior to it, the Department of the Interior, under authority
of section 8(k) of the Outer Continental Shelf Lands Act (OCSLA) has
planned for leasing and eventually exploiting mineral resources on the
OCS and in the EEZ. Prior to creation of the EEZ, Interior prepared
two draft environmental impact statements: Proposed Hard Mineral Mining
Leasing and Operating Regulations, 1973; and Proposed Polymetallic
Sulfide Minerals Lease Offering, Gorda Ridge Area, 1983 (begun before
EEZ Proclamation).

As the DOI looks toward mineral leasing on the OCS/EEZ, evaluation
of possible environmental effects becomes an important policy decision
criterion. Given the ongoing public debate over the nature of the
environmental effects from the offshore oil and gas program and
increasing public awareness and concern for the health of the marine
environment, solid mineral mining offshore will attract considerable
attention on environmental issues. It is essential that solid mineral
progam administrators and decision-makers have available the most
complete picture on the potential and likely environmental affects of
marine mining. As well, complete information arms the interested
public with the facts and not misperceptions based on incomplete
knowledge.

This paper considers how the findings of the recently concluded
OCSLA Section 20(e) Report (OCS Report, 1988) on the cumulative
environmental effects of the Department's OCS oil and gas program can
be used to help evaluate and understand the likely environmental
effects from mining in the oceans.

2. IMPLICATIONS OF THE OCSLA CUMULATIVE EFFECTS REPORT

The 20(e) Report is an annual report to Congress on the cumulative
effects on the environment of the activities conducted under the OCSLA.

D. A. Ardus and M. A. Champ (eds.), Ocean Resources, Vol. I, 321–330.

To date those activities have been limited to oil and gas exploration and development (Sulphur mining on the OCS has been limited to a few operations and its impacts are not differentiated from oil and gas drilling because the impacts of the Frasch solution mining process are so similar). Last year, the DOI published its first cumulative effects report, a retrospective on the offshore program from 1954 to 1987. Generally, the 20(e) Report concluded that except for loss of wetlands in the Mississippi delta, the cumulative environmental effects attributable to the OCS oil and gas program have been minimal.

This finding is important because it is so different from the projected assessment of environmental effects contained in the environmental impact statements (EISs) for offshore oil and gas leasing and development. Also, public perception of Federally proposed or licenced projects is shaped largely by what is said in the EIS.

Environmental impact Statements tend to focus on short term, acute environmental effects that are well known and immediately observable. They focus on potential impacts which overshadow assessments for likely effects. Further, when essential information is unknown or uncertain, the required "worst case" analysis may over predict adverse environmental consequences. This situation may exist for EISs prepared for leasing marine solid minerals, especially at the leasing stage when so much information is not available or must be surmised, such as the extent or composition of ore bodies, use of untried or unproven mining technologies and onshore beneficiation processes.

Thus, a solid mineral leasing EIS may cloud rather than clarify the impacts that may result. For example, in the DOI proposed lease sale EIS for cobalt-rich crusts, the hypothetical mining scenario used for EIS analysis purposes is estimated to be three times larger than may be economically feasible. That scenario may be appropriate for NEPA analysis, but amplifies potential levels of environmental impacts (Draft EIS, 1988).

In contrast, the 20(e) Report examines only the observed effects of the OCS oil and gas program as documented in the scientific literature. The perspective of the 30-plus year impact of OCS development separates the short term/near field perturbations from the long term/far field environmental impacts which have been found to be largely insignificant. As a model for comparing projected effects and observed effects, the 20(e) Report can assist public policy decisions that must balance the impacts of marine mining with environmental protection.

3.   COMPARISON OF OIL AND GAS WITH MINING ACTIVITIES

The marine environments for historic oil and gas development are comparable to those where marine mining may occur. The OCS/EEZ mineral resources being examined by the DOI are: gold in Norton Sound in Alaska, polymetallic sulfide deposits and "black" sands off Washington and Oregon, phosphate deposits off the southern California, North Carolina, and Georgia coasts, cobalt-rich crusts around the Hawaiian Islands and Johnston Island, ilmenite sands off the mid-Atlantic coast, and sand and gravel construction materials wherever they occur (Federal

Register, 1988). With the exception of the Hawaiian Islands/Johnston Island area, the environments for these solid minerals are substantially the same or similar to the present OCS leasing and operating areas for oil and gas.

Offshore oil and gas operations are well developed and rely on "standardized" technology. No such standard exists for marine mining because of the variety of target mineral deposits. Four basic methods for marine mining are

(1) surficial scraping consisting of suction dredging, drag line dredging, and crustal miners for unconsolidated, nodule, and crustal deposits;

(2) excavation using a clam shell bucket, bucket ladder or bucket wheel dredge, cutterhead suction dredging, and drilling and blasting for thick or buried deposits and for massive or vein deposits;

(3) fluidizing for sub-seabed mining of unconsolidated deposits using a slurry, or using a leaching solution for bedded, massive, and vein deposits; and

(4) tunneling beneath the seabed from shore or from artificial islands (OCS Report, 1987).

Despite the variety of marine mining methods, they generate environmental impacts that are analogous to those for offshore oil and gas activities. The comparison facilitates extrapolation of cumulative environmental effects from oil and gas to mining. Table 1 shows a side by side comparison of the steps in fluid and solid mineral operations.

Geologic and geophysical (G&G) exploration methods are virtually identical, using seismic, sonar, magnetic and gravimetric remote sensing. Physical sampling includes shallow coring, and seabed and water column sampling. Exploration drilling for oil and gas is the same as deep drilling to identify solid mineral targets and for ore body delineation. Except for solution mining, oil and gas development operations are quite different from marine mining operations even though, as shown on Table 1, many of the supporting activities and impact generating factors are the same.

4.   ENVIRONMENTAL EFFECTS

The 20(e) Report provides a thorough analysis of environmental effects of the OCS oil and gas program. For this paper, the focus is on impacts to several resources: water quality, air quality, fish, benthic and marine biota, and marine mammals. These resources are emblematic of the key environmental concerns for both offshore oil and gas and mining operations.

TABLE 1. Comparison of Offshore Oil and Gas and Marine Mining
Activities.

Oil and Gas Related	Mining Related

Geological and Geophysical
  Exploration
    Remote Sensing
    Geological Sampling
	Geological and Geophysical
	  Exploration
	    Remote Sensing
	    Geological Sampling

Exploration Drilling
    Setting Structures/Anchors
    Drilling
 	
    Support Activity
    Discharges
    Emissions
	Test Mining
	    Setting Structures/Anchors
	
	    Mining
	    Support Activity
	    Discharges
	    Emissions

Construction
    Platform Setting
    Pipeline Construction
	Construction
	    Fixed Structure

Development Drilling/Production
    Drilling
    Hydrocarbon Production
    Processing/Separation
    Oil/Gas Transport
    Support Activity
    Discharges
    Emissions
    Offshore Processing/Refining
    Platform Removal
	Mining
	    Mining
	    Ore Production
	    Processing/Beneficiation
	    Ore Transport
	    Support Activity
	    Discharges
	    Emissions
	    Onshore Processing/Refining

4.1. Water Quality

Oil and gas activities affect water quality through routine discharges
of drilling muds and cuttings, wastes from vessels and platforms,
produced formation waters, and accidental spills of crude oil or fuel.
Dredging for platform installation and pipelaying resuspends sediments
and creates turbidity. Excavation and dredging activities for mining
also disturb the seabed causing sediment plumes. Depending upon how
the mined material is lifted to the surface, suspended sediment may be
released anywhere in the water column. Mining vessel discharges
include surface discharge of water and or tailings, wastes from
vessels, and accidental releases of fuel. All discharges to OCS/EEZ
waters are regulated under the Clean Water Act.

The discharge of drilling muds and cuttings creates a surface and
or benthic turbidity plume, depending on point of discharge. Dilution
of the discharge to low concentrations is rapid, with solid
concentrations reduced to 1000 ppm within 2 minutes of discharge and 10
ppm within 1 hour (Neff, 1981 and 1985). Typically the effected zone
down current of discharge extends 1000-2000 m (ECOMAR, 1978 and 1983).

Muds, produced formation waters, and other waste discharges pose only
near field water quality concerns. Muds and formation waters do
contain trace quantities of arsenic and metals, however, monitoring
studies in the Gulf of Mexico (GOM) indicate that pollutants are
rapidly reduced to background levels (Middleditch, B.S., 1981). Long
term, chronic environmental effects of these discharges have not been
observed. Because of the high potential for adverse impacts on marine
ecosystems, this remains a subject for further study.

TABLE 2. Comparison of Gulf of Mexico Muds and Cuttings Discharges
with Projected Offshore Mining Discharges

Offshore Activity	Annual Discharge[a]	
	Metric Tonnes	Cubic Metres
GOM Oil and Gas Drilling[b]	1.3×10^6	9.2×10^5
Proposed OCS Gold Mining, AK	-	22.5×10^5
Proposed EEZ Cobalt Crust Mining, HA	1.5×10^6	-
Proposed Polymetallic Sulfide Mining, Gorda Ridge	-	7.2×10^5
Proposed Deep Seabed Mining	-	59.8×10^5

[a] - Working Days: GOM Drilling-365 days; Gold Mining-150; Cobalt
    Crust Mining-206; Polymetallic Sulfide and Deep Seabed
    Mining-each 300.

[b] - Approximate conversions based on annual discharge of 5.8
    million barrels of muds and cuttings.

The volume of muds and cuttings from oil and gas as compared with
sediment inputs from mining sources are as shown in Table 2. Annual
discharges of muds and cuttings into GOM are estimated to be 5.8
million barrels. Hypothetical mining scenarios in EISs project
comparable annual discharges for operations, although the mining
discharges would be geographically and temporally more concentrated.
Though the magnitude of discharges from a mining operation may be
greater, the same settling and dilution factors that minimise the
effect of oil and gas related discharges apply. As with any activity
that resuspends sediments, down current turbidity may be detected over
many kilometres, however, as observed for dredging operations, direct
effects are limited to the immediate area of operation (US Army Corps
of Engineers, 1978). Indirect effects such as nutrient or trace metal
enrichment, increased biological or chemical oxygen demand and reduced
penetration of sunlight may be experienced proximate to any mining
operation.

        Accidental releases of oil into the marine environment have been
well studied. The 20(e) report considers oil spill effects on water
quality and potential impact on birds, marine mammals, fish, and
coastal resources. The effects of large spills may be dramatic and can
kill large numbers of sea birds and shore birds if they come into
contact with oil, but the effects are highly localized and generally
short lived. Oil in the water undergoes weathering and organic
decomposition diminishing the oil over time. The Santa Barbara Channel
spill of 1969 had few observable effects noted after less than 2 years
(Straughan, 1971). Unlike blowouts during oil and gas operations,
spills from marine mining activity would be from fuels, mostly during
transfer, but also from vessel loss. These effects, as noted for the
OCS oil and gas program would be minimal.

4.2. Air Quality

Air emissions come from power generation and engine sources on or used
to support an offshore drilling platform or mining vessel. Principle
emissions are nitrous oxides and residual organic compounds. During
activities like G&G, exploration, and test mining, temporary pollutant
emissions may have little effect onshore. The effects of air emissions
from OCS sources are dependent upon the location of the activity and
prevailing meteorological and onshore air quality conditions. Because
most oil and gas operations are so far from shore, OCS emissions,
except offshore California, have not been demonstrated to adversely
affect onshore air quality. Pollutant emissions from marine mining
sources are expected to be qualitatively and quantitatively similar to
oil and gas related sources. The MMS regulates emissions from OCS
operations, including mining, to ensure that onshore National ambient
air quality standards are protected.

4.3. Fishery Resources

Two aspects are the fish as a resource and fishing as an economic
activity. Standing stocks of the resource are affected by turbidity,
pollutant loading, and physical disturbance, particularly to the extent
recruitment may be reduced. Marine mineral activities may interfere
with fishing activities and compete for space at sea and in port.
        Direct effects of oil or turbidity can be avoided by most fish
owing to their motility. Indirect effects include damage to eggs,
larvae, and juveniles; sublethal uptake of hydrocarbons and pollutants;
loss of prey; loss of habitat; or reduced reproductive success. It is
virtually impossible to isolate the effects of offshore oil and gas on
mobile populations like fish, but offshore platforms do support reef-
like communities of fish with little evidence of tainting or adverse
effect. Oil and turbidity can reduce recruitment but no such effects
have been identified.
        Of recent interest has been the potential for G&G surveys to
reduce catchability of fish and damage fish eggs and larvae. Limited
studies have shown that long duration, spatially concentrated use of
seismic energy sources can disturb the spatial distribution of fish in
the water column and reduce catchability (Greeneridge Sciences, 1985

and Battele Marine Res. Lab. & BBN Labs., 1987). Ongoing studies of
seismic energy source (Personal communications, 1988). Despite over
8000 permits issued for offshore oil and gas G&G activities and
millions of line miles of seismic surveys having been compiled, no
cumulative adverse effects fishery resources have been detected.

Relative to commercial fishing, space use conflicts between
fishermen and vessel operators have occurred with entanglement or
severing of net and trap lines. Coordination efforts between the two
industries have helped avoid most vessel conflicts. Offshore drilling
rigs and platforms also can preclude access to fishing grounds and
debris from offshore activities can damage fishing gear. Except for
fixed platforms, the displacement effect is temporary and has no effect
on populations of commercial fish species. It is expected that there
has been some loss of individual income through lost catch opportunity
or gear loss and increased cost of port space.

Similar effects would be expected from mining activities.
Disturbance of large areas of the seabed by mining would alter
habitats, which could affect feeding and spawning. An important impact
that cannot be assessed in advance is the uptake of heavy metals which
might be released by mining or tailings discharge. Trophic effects,
from uptake by plankton through human consumption, would be a major
concern.

4.4. Benthos

Sources of impact to benthos are physical removal of the seabed and
effects of turbidity and contaminated discharges, all of which result
from both oil and gas or mining operations.

Though oil and gas activities disturb the seabed and bury adjacent
benthic habitat, the areal extent of this effect is small, generally
less than 1000 square meters (U.S. Dept. of the Interior, 1988).
Moreover, muds and cuttings may be recolonized and, although the new
benthic community may be different from the natural one, opportunistic
fauna may demonstrate great success (Clark, 1988). Despite long
standing and wide ranging oil and gas related discharge activities in
the GOM, no cumulative adverse effects on benthos have been detected.

Elimination of the benthic substrate is a singular property of
most mining methods. Compared with individual oil and gas activities,
the area affected by removal and by smothering from deposition of
suspended sediments during mining is greater by several orders of
magnitude. However, referencing Table 2, while the local impact to
benthos would be severe, the overall effect on the marine environment
and productivity may be insignificant.

4.5 Marine Mammals

Impacts to marine mammals, many of which are endangered or threatened
species, may come from contact with oil, loss of feeding areas and
trophic effects due to uptake of heavy metals, and noise effects.

Oil in the water can be inhaled by marine mammals surfacing to
breathe, ingested while grooming or from contaminated prey, may cause
skin irritations, and, as in the case of sea otters, can reduce the

insulating properties of fur. Despite potential impacts, there has
been little if any effect on populations of marine mammals from
offshore oil and gas activities. Even the Santa Barbara Channel spill,
occurring at the height of the gray whale migration, had no effect on
these animals (Brownell, 1971). With smaller quantities of oil at risk
of spillage from marine mining, minimal effects would be expected.

Many marine mammals are migratory and wide ranging and can avoid
the effects of turbidity plumes. Food prey is unlikely to be present
in plumes further reducing the likelihood of contact. Because whales
and some other marine mammals are long lived, there is a chance of
accumulating drilling mud or formation water derived metals. There
were no findings documenting contamination of marine mammals in the
20(e) Report, but this may not be representative of likely effects from
solid mineral mining. Release of ore derived heavy metals from mining
activity has been a controversial issue in solid mineral EISs (USDOI,
MMS, 1988). Recent strandings of bottlenose dolphin along the Atlantic
coast and the harbor seal mortality in Northern Europe point to high
levels of environmental stress, from all sources, and indicate a need
for careful monitoring of pollutant loading in the marine environment.

Noise is a ubiquitous effect from any offshore operation,
potentially impacting endangered whales and marine mammals by altering
behaviour. For oil and gas offshore Alaska, the issue has focused on
how operational noise may affect whale migrations, feeding behaviour,
breeding, and calving. Location and timing are critical aspects for
noise effects. Some exploratory operations in the Beaufort Sea that
are in the path of migrating endangered bowhead whales have been
seasonally restricted (historically) so that the whales do not alter
their migration path as a noise avoidance response. Despite the
sensitivity of some animals to noise, the 20(e) Report notes no long
term adverse effects from noise on either populations or species
behaviour. Most behavioural responses are elicited at such close range
that noise effects can be avoided by animals. Mining activities
located away from known migratory pathways and calving or feeding
grounds are unlikely to adversely affect marine mammal populations
although individual transient animals near mining sites may be startled
or show avoidance behaviour. However, because all oil and gas or
mining activities are so site specific, it may be difficult to
eliminate all noise effects if activities are proposed in sensitive
habitats.

5. CONCLUSION

Marine mining operations are expected to generate many of the same
impacts as the current OCS oil and gas program. By examining the
effects of more than 30 years of OCS activity documented in the 20(e)
Cumulative Environmental Effects Report, public policy officials and
the public may be able to better understand and anticipate the
environmental effects of marine mining in the EEZ.

6.   REFERENCES

Battele Marine Research Laboratory and BBN Laboratories (1987)
     Effects of Sounds from a Geophysical Survey Device on Fishing
     Success, prepared for MMS Contract No. 14-12-001-30273.

Brownell, R.L. Jr. (1971) 'Whales, Dolphins and Oil Pollution' in
     Straughan, D. (ed.), Biological and Oceanographical Survey of
     the Santa Barbara Channel Oil Spill, 1969-1970, pp.287-306.

Clark, R.B. (1988) 'Impact of Offshore Oil Operations in the North
     Sea' in Proceedings of Oceans' 88 Conference, Vol. 1,
     pp.184-187.

ECOMAR (1978) Tanner Bank Mud and Cuttings Study, prepared for
     Shell Oil Company, California, and

ECOMAR (1983) Mud Dispersion Study: Norton Sound COST Well No. 2,
     Prepared for ARCO, Alaska.

Greeneridge Sciences (1985), Pilot Study on the Dispersal of
     Rockfish by Seismic Exploration Acoustic Signals, prepared for
     Joint Commercial Fishing/Petroleum Exploration Industries
     Project.

Middleditch, B.S. (1981) Environmental Effects of Offshore Oil
     Production: Buccaneer Gas and Oil Field Study, Plenum Press, New
     York and London, 446p.

Neff, J.F. (1981) 'Fate and Biological Effects of Oil Well
     Drilling Fluids in the Marine Environment', Final Technical
     Report to U.S. Environmental Protection Agency.

Neff, J.F. (1985) 'Biological Effects of Drilling Fluids, Drill
     Cuttings and Produced Waters' in Boesch, D.F. and Rabalais, N.N.
     (eds.), The Long Term Effects of Offshore Oil and Gas
     Development, An Assessment and Research Strategy, Report to
     NOAA.

Personal communications on preliminary results of a study on the
     effects of seismic survey energy pulses on eggs and larvae of
     the dungeness crab, in preparation for the Joint Commercial
     Fishing/Petroleum Exploration Industries Project (1988).

Straughan, D. (1971) 'Biological and Oceanographic Survey of the
     Santa Barbara Channel Oil Spill', 1969-1970, Sea Grant
     Publication,   No. 2, University of Southern California.

U.S. Army Corps of Engineers (1978) Effects of Dredging and
     Disposal on Aquatic Organisms, Waterways Experiment Station
     Technical Report DS-78-5, 41p.

U.S. Department of the Interior (1973) Proposed Hard Mineral
    Mining Leading and Operating Regulations.

USDOI (1983) Proposed Polymetallic Sulfide Minerals Lease
    Offering, Gorda Ridge Area.

USDOI, Minerals Management Service (1987) Environmental Effects
    Overview: Marine Mining on the Outer Continental Shelf, OCS
    Report, MMS 87-0035, 66p.

USDOI, MMS and Hawaii Department of Planning and Economic
    Development (1988) Draft EIS, Proposed Marine Mineral Lease Sale
    in the Hawaiian Archipelago and Johnston Island Exclusive
    Economic Zone, 344p. (p.37).

USDOI, MMS (1988) 30 CFR Part 282: Operations in the Outer
    Continental Shelf for Minerals Other than Oil, Gas and Sulphur;
    Proposed Rule, Federal Register, Vol. 53, No. 160, pg.31442,
    August 18th.

USDOI, MMS (1988), Oil and Gas Program: Cumulative Effects, OCS
    Report 88-0005.

USDOI, MMS (1988), Draft EIS, OCS Mining Program: Norton Sound
    Lease Sale, MMS Report 88-0082, 357p.